CONTENTS

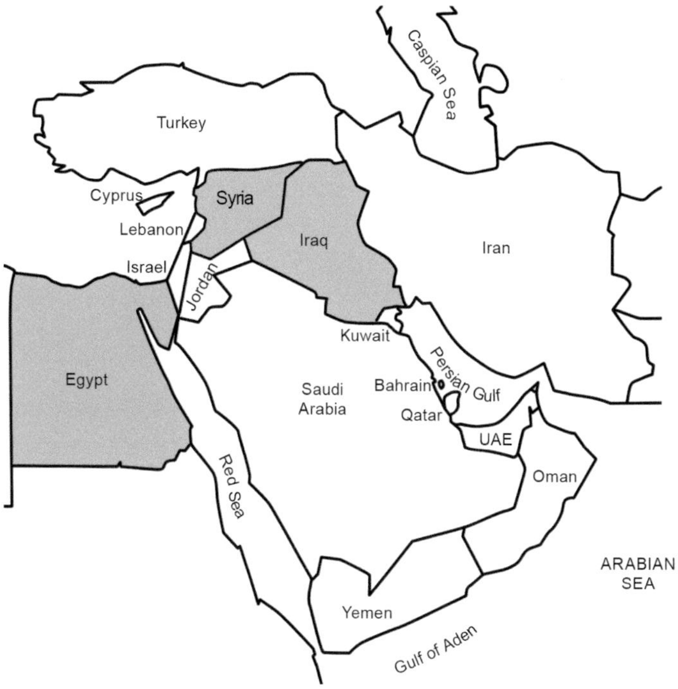

NOTE

In order to simplify the use of this book, all names, locations and geographic designations are as provided in The Times World Atlas, or other traditionally accepted major sources of reference, as of the time of described events. Similarly, Arabic names are romanised and transcripted rather than transliterated. For example: the definite article al- before words starting with 'sun letters' is given as pronounced instead of simply as al- (which is the usual practice for non-Arabic speakers in most English-language literature and media). Instead of using the diacritical marks to represent the letter *'ayn*, double a is used, while names like 'Faisal/Feisal' are spelled as 'Faysal'.

Helion & Company Limited
Unit 8 Amherst Business Centre, Budbrooke Road, Warwick CV34 5WE, England
Tel. 01926 499 619
Email: info@helion.co.uk Website: www.helion.co.uk Twitter: @helionbooks Visit our blog http://blog.helion.co.uk/

Published by Helion & Company 2021
Designed and typeset by Farr out Publications, Wokingham, Berkshire
Cover designed by Paul Hewitt, Battlefield Design (www.battlefield-design.co.uk)

ISBN 978-1-913336-36-3

British Library Cataloguing-in-Publication Data.
A catalogue record for this book is available from the British Library.

For details of other military history titles published by Helion & Company Limited contact the above address, or visit our website: http://www.helion.co.uk. We always welcome receiving book proposals from prospective authors.

ABBREVIATIONS

AB	air base
ADC	Air Defence Command (Egypt)
AMAN	Agaf HaModin (Military Intelligence Directorate, Israel)
An	Antonov (the design bureau led by Oleg Antonov)
ASCC	Air Standardisation Coordinating Committee
BAC	British Airspace Corporation (later BAe Warton)
BAe	British Aerospace
BAF	Bangladesh Air Force
CAP	combat air patrol
CAS	close air support
CGS	Chief of General Staff
CIA	Central Intelligence Agency (USA)
CO	commanding officer
COIN	counterinsurgency
DHC	de Havilland Canada
DIA	Defence Intelligence Agency (USA)
DM1	Main Directorate of Military Intelligence (Egypt)
DMI	Directorate of Military Industries (Egypt & Iraq)
DMZ	de-militarised zone
EAF	Egyptian Air Force (1952-1958 & 1972 until today)
FAC	forward air controller
FCS	fire-control system
GenStab	General Staff of the Soviet Armed Forces
GMID	General Military Intelligence Directorate (Iraq)
GRU	Glavnoye razvedyvate'lnoye upravleniye (Main Intelligence Directorate of the Soviet Armed Forces)
HAWK	Homing-All-The-Way-Killer (US-made SAM)
IAF	Indian Air Force
IAP	international airport
IDF	Israel Defence Force
IDF/AF	Israel Defence Force/Air Force
Il	Ilyushin (the design bureau led by Sergey Vladimirovich Ilyushin, also known as OKB-39)
INA	Iraq News Agency
IPC	Iraq Petroleum Company (1929-1961)
IrAAC	Iraq Army Aviation Corps
IrAF	Iraq Air Force (1958-2003)
IrAF-ID	Iraq Air Force – Intelligence Department
IRGC	Islamic Revolutionary Guards Corps (Iran)
KAF	Kuwait Air Force
LAAF	Libyan Arab Air Force
Luftwaffe	German Air Force
MiG	Mikoyan i Gurevich (the design bureau led by Artyom Ivanovich Mikoyan and Mikhail Iosifovich Gurevich, also known as OKB-155 or MMZ 'Zenit')
NATO	North Atlantic Treaty Organisation
NWC	National Water Carrier (Israel)
ONI	Office of Naval Intelligence (US Navy)
OPEC	Organization of Petroleum Exporting Countries
PAF	Pakistan Air Force
RAF	Royal Air Force (United Kingdom, since 1918)
RBAF	Royal Bahrain Air Force
RCC	Revolutionary Command Council (Egypt)
RSAF	Royal Saudi Air Force
RWR	radar warning receiver
SAM	surface-to-air missile
SEATO	Southeast Asia Treaty Organisation
SNCASO	Sud-Oest Aviation (France)
Su	Sukhoi (the design bureau led by Pavel Ossipovich Sukhoi, also known as OKB-51)
SyAAF	Syrian Arab Air Force
TPC	Turkish Petroleum Company (1912-1925; subsequently IPC)
UAEAF	United Arab Emirates Air Force
UAR	United Arab Republic (Egypt and Syria from 1958 until 1961, and Egypt until 1972)
UARAF	United Arab Republic Air Force (EAF and SyAAF from 1958 until 1961, and EAF until 1972)
UNRWA	United Nations Relief and Works Agency for Palestine Refugees in the Near East
USAF	United States Air Force
USN	United States Navy
USSR	Union of Soviet Socialist Republics (also 'Soviet Union')
VKS	*Vozdushno-kosmicheskiye sily* (Russian Aerospace Forces, since 2015)
Yak	Yakovlev (the design bureau led by A. S. Yakovlev, also known as OKB-115)

INTRODUCTION AND ACKNOWLEDGEMENTS

On 1 November 1911, Lieutenant Giulio Gavotti, the pilot of an Etrich Taube reconnaissance aircraft of the Italian armed forces, dropped three small bombs upon 'Turkish positions' outside Ein Zara, then a minor town east of Tripoli, the capital of the Ottoman province of Tripolitania. This action is remebered as the world's first ever bombing attack from the air. While it remains unknown if the 1.5 kilogram bombs dropped by Gavotti caused any kind of casualties, it is known that his actual target was a group of Arab tribesmen allied with the Ottoman Empire: 'Arabs' thus became the first ever target of an air strike. Over the last 110 years, they also became perhaps the most-frequently-targeted ethnic group. Those in Libya remained on the receiving end of Italian air strikes well into the 1930s; those in the Levant on the receiving end of countless British and French air strikes during the same period. Indeed, the Arabs of Iraq in the 1920s almost became the first targets of chemical weapons deployed from the air, and were certainly the first to find themselves on the receiving end of what used to be known as 'air policing', and has been described in very different terminology ever since: an intentional aerial assault upon civilians with the aim of forcing them into submission. Air policing included such practices as what became known as 'fire-bombing' during the Second World War, and 'double-tap air strikes' in more modern times. Since at least 1939, the centrepiece of armed conflicts fought in the Middle East has been the 'Arab-Israeli dispute' and most of the resulting armed conflicts have seen extensive use of air power – if they were not dominated by it – often on a scale rivalling earlier or contemporary conflicts elsewhere, and frequently including 'terrorist' air strikes upon civilian populations. This was true for the Suez War of 1956, the June 1967 Arab-Israeli War, the so-called War of Attrition of the 1967-1973 period, the October 1973 Arab-Israeli War, but especially for the three Israeli invasions of Lebanon in the 1970s and the 1980s; for repeated 'limited operations' of the Israel Defence Force/Air Force (IDF/AF); since 2004 re-designated as the Israeli Air and Space Force (IASF) against Palestinian militants abroad

(Tunis 1985), and in the Gaza Strip (2006, 2008-2009, 2012, 2014, 2018-2020), and the Israeli operations against Syria (at least since 2007, probably since 2001). Meanwhile, air power has been applied in a myriad of other wars involving Arab states – some by foreign powers, and others fought between different states in the Middle East. The most prominent examples would be the Yemen Civil War of 1962-1970; several military coup attempts in Iraq in the 1960s; the Egyptian-Libyan War of 1977; the war between North and South Yemen in 1979; the Iran-Iraq War of 1980-1988 (also known as the First Persian Gulf War); the war for the liberation of Kuwait fought between a US-led coalition and Iraq, in 1990-1991 (or the Second Persian Gulf War), the 'air policing' of Iraq by the USA, Great Britain and France between 1991-2003, and the US-invasion of Iraq and the resulting 'civil' war since 2003. Another, and perhaps the most tragic application of air power in the last few decades has taken place in Syria since 2011, where a series of ruthless aerial assaults on the civilian population – frequently including deployment of chemical weapons – by the Syrian Arab Air Force (SyAAF) and since 2015, the Russian Aerospace Force (VKS), converted nearly 80 percent of the population into homeless refugees, and drove over 60 percent out of the country, in turn causing a major refugee crisis, strongly felt well outside its geographic limits. Considering such a 'rich' history of aerial warfare in the Middle East, it is rather unsurprising that at the time these lines are written, air forces of the United States, Iraq and several allies continue 'mopping up' operations against extremists of the so-called 'Islamic State' (IS) in Iraq and sometimes in Syria, too; US unmanned aerial vehicles are deployed for sporadic attacks on IS elements in Iraq, Syria and Yemen; the Turkish Air Force is bombing Kurdish militants in northern Iraq; while the Kuwait Air Force, Royal Bahrain Air Force (RBAF) and Royal Saudi Air Force (RSAF) – all supported by tanker aircraft of the US Air Force (USAF) – are taking part in an incredibly complex war fought in Yemen for decades. This is not to forget that, following four years of intense deployment on the side of the Saudi-led coalition in Yemen, time and again, the United Arab Emirates Air Force (UAEAF) is involved in the similarly complex 'civil wars' in Libya and Ethiopia.

Ironically, despite such a widespread and – often enough – intensive application of airpower, and thus the high degree of influence that this means of warfare exercises upon the contemporary history of the Middle East, the mass of related publications almost entirely ignores the local air forces. The number of specialised studies about these published over the last seven decades is lower than the number of the publications about the Israeli or major Western air forces published almost every single year. Indeed, regardless of what Arab-Israeli conflict they cover, the mass of related publications entirely concentrate on the IDF/AF. An unsurprising, but frequently grotesque result is that there is still very little understanding even about markings applied on the aircraft of the different Arab air forces, not to mention precise details of their doctrine, strategy and tactics, organisation, operational history, equipment – or even the number of aircraft acquired, and their appearance. Arguably, the Arab states have released relatively little official documentation – or even the official viewpoints of their governments – about their air forces: or if they have, usually only in the Arabic language. Not only for these reasons, but for reasons related to little more than hatred and propaganda, next to nothing of this material has ever been translated into any of the Western languages, even less published, and thus there has been next to no opportunity for professionals and enthusiasts alike to adequately study the application of Arab air power, and related experiences. Nowhere is this as valid as for the series of Arab-Israeli wars fought between the 1950s and the 1970s, during which Soviet-designed combat aircraft formed the backbone of several Arab air forces.

The reality is that wars are always – at least – two-sided affairs: no matter to what degree one side claims to be party to knowledge, indeed 'only telling the truth', and that the other side 'is always lying', not one actor is capable of providing a complete picture or an understanding of historical 'truth' and even less so when emotionally involved. Unavailingly, completely ignoring one party, while entirely concentrating on the other, indeed considering both the official and unofficial storyline of one party as something like 'the only truth', cannot but result in gross misinterpretation and a mass of mistakes.

While acting as a sequel to the *Air Power and the Arab World* mini-series, and a complimentary source of reference for a number of other books published in Helion's Middle East@War series (for details, see the bibliography), the principal purpose of this project can thus be described as an 'upgrade of an upgrade': this is, essentially, the third generation of publications on this topic by the same core-team of authors. Th primary aim is inclusiveness and emphasising contextualisation while reconstructing the operational history of the Arab air forces equipped with military aircraft of designs from the Union of Soviet Socialist Republics (USSR/Soviet Union) – including those of the Antonov, Ilyushin, Mikoyan i Gurevich, Sukhoi and Yakovlev design bureaus – between the mid-1950s and the mid-1970s. For this purpose, we have reviewed not only our own publications from the last 20 years, but also dozens of related Western and Israeli publications, have upgraded everything with newly acquired information and photographs, wherever necessary and possible and then focused the narrative on the air forces operating Soviet-made combat aircraft.

The Israeli and Western points of view and impressions of the Arab-Israeli conflicts are widely publicised and should be well-known even to less experienced readers. Correspondingly, Israel was supposedly created by 'survivors of the Holocaust' and has ever since been a tiny little country surrounded by blood-thirsty enemies determined to destroy it at the first opportunity. Western accounts tend to look at the Arab-Israeli conflict through the prism of the Cold War – the major stand-off between the USSR-dominated Eastern Europe and the US-allied West lasting from the 1940s until the early 1990s. According to the lines represented in the mass of such publications, the Arabs were – and remain – aggressive and extremists, and it is them to blame for not only everything that went wrong, but indeed for every armed conflict fought in the Middle East at least since 1918. Therefore, and while having no axe to grind, in this work we intentionally emphasise the Arab point of view, and try to explain the Arab reasoning. After all, various Arab statesmen, governments, air forces and other armed services have had their own reasons for purchasing Soviet armaments. Although these reasons have changed several times over the decades, all were very local by their nature. In turn, these, *local* reasons have not only conditioned such technical details as the type, equipment and armament of the aircraft in question, the quality of training their crews have received, or the way they were deployed in combat, but even their camouflage colours and markings.

The research that led to this study began over 50 years ago and was run with the help of countless participants and eyewitnesses over that time. Thus, there are many people who have kindly shared – often intriguing – personal stories, and to whom the authors would like to express their gratitude. Foremost are the officers and pilots of the Arab air forces including late Air Marshal Alaa el-Din Barakat (EAF), late Air Marshal Mustafa Shalabi el-Hinnawy

(EAF), Air Marshal Farouq el-Ghazzawy (EAF, ret.), Air Marshal Badr Domair (EAF, ret.), late Lieutenant-General Arif Abd ar-Razzaq (IrAF), late Air Marshal Tahir Zaki (EAF), Air Vice-Marshal Ahmed Abbas (EAF, ret.), Major-General Alwan Hassan al-Abossi (IrAF, ret.), Major-General Ahabadin Ayman (ADC, ret.), Air Vice-Marshal Qadri Abd el-Hamid (EAF, ret.), Air Vice-Marshal Zia el-Hefnawi (EAF, ret.), Air Vice-Marshal Mamdouh Heshmat (EAF, ret.), Air Vice-Marshal Reda el-Iraqi (EAF, ret.), Major-General Hussein el-Kfass (EAF, ret.), Air Vice-Marshal Ahmed Kilany (EAF, ret.), Air Vice-Marshal Mustafa Nabil al-Masri (EAF, ret.), Air Vice-Marshal Abd al-Moneim Mikaati (EAF, ret.), Air Vice-Marshal Samir Aziz Mikhail (EAF, ret.), Major-General Nassr Moussa (EAF, ret.), Major-General Mohammed Naji (IrAF, ret.), late Air Vice-Marshal Mohammed Abdel-Moneim Zaki Okasha (EAF), Major-General Salim Saffar (IrAF, ret.), Air Vice-Marshal Siad Shalash (EAF, ret.), Air Vice-Marshal Sa'ad ad-Din Sherif (EAF, ret.), Air Vice-Marshal Nabil el-Shuwakri/Shoukry (EAF, ret.), Air Vice-Marshal Mamdouh Taliba (EAF, ret.), Major-General Ihsan Shurdom (RJAF, ret.), Major-General Medhat Zaki (EAF, ret.), Major-General Ahmed Yusuf (EAF, ret.), Air Vice-Marshal Tahsin Zaki (EAF/ADC, ret.), late Brigadier-General Ahmad Sadik Rushdie al-Astrabadi (IrAF), Brigadier-General Faysal Abdul Mohsen (IrAF, ret.), Brigadier-General Farouk Abdeen (RJAF, ret.), Air Commodore Tamim Fahmi Abdullah (EAF, ret.), late Air Commodore Gabr Ali Gabr (EAF, ret.), Air Commodore Fikry el-Gahramy (EAF, ret.), Air Commodore Fikry el-Gindy (EAF, ret.), late Air Commodore Mustafa Mohammed Hassan (EAF), late Air Commodore Mustafa Hafez (EAF), Air Commodore Fuad Kamal (EAF, ret.), Air Commodore Abdel Moneim el-Tawil (EAF, ret.), Air Commodore Ibrahim Gazerine (EAF, ret.), Group Captain Kapil Bhargava (IAF, ret.), late Group Captain Saif-ul-Azam (PAF/BAF), Wing Commander Talaat Louca (EAF, ret.), Wing Commander Usama Sidqi (EAF, ret.), Wing Commander Kamal Zaki (EAF, ret.), Squadron Leader Wagdi Hafez (EAF, ret.), Captain Abdelmajid Tayyari (LAAF, ret.).

Our thanks go to Mr. Tarek el-Shennawy (pilot of Egypt Air and son of late Air Vice-Marshal Abdel Moneim el-Shennawy), Mr. Ahmad Keraidy (pilot of Egypt Air and son of late Air Vice-Marshal Abdel Wahhab el-Keraidy), Mr. Gamal Khalifa (son of late Air Commodore Nazih Khalifa), Mr. Tarek O Sidki, Mrs. Patricia Salti (leading historian of the Royal Jordanian Air Force), Mrs. Leila, the late Mrs. Khouda, and the late Mrs Mona Tewfik for permission to use their family archives. We would also like to express our special thanks to Martin Smisek from the Czech Republic for working himself through the Czech National Archive for many years; to Nour Bardai, Dr. Abdallah Emran, and Sherif Sharmi and Hebatallah Ghanem from Egypt; to Ali Tobchi from Iraq, and Albert Grandolini from France for running additional interviews and faithfully providing plenty of precious bits and pieces of information over the years. We wish to thank Ass'aad Dib in Lebanon for his translations of Ali Muhammad Labib's history of the Egyptian Air Force, to Milos Sipos from Slovakia for his kind help with research about the Iraqi Air Force, and Lon Nordeen in the USA, for permission to use some of his research including interviews with a number of Egyptian participants. Last, but not least, our thanks go to Farzin Nadimi for his research in the archives of the Air Ministry in Great Britain, and to Hicham Honeini from Lebanon for his patience and kind help with translations of various publications and documentation from Arabic.

1
BACKGROUND

With the Middle East being fully exposed to some of the worst applications of imperialism and its machinations since at least the 1910s, then to the colonisation of Palestine by Jewish immigrants, and subsequently to widespread subversion and subjugation to Western economic interests, relations between the majority of the local population and Western powers were deeply troubled long before governments of (in alphabetic order) Algeria, Egypt, Iraq, Libya, Morocco, Syria and Yemen took the decision to obtain Soviet-designed military aircraft.[1] Relations between 'Arabs' and Western powers became deeply troubled almost as soon as they were established, in the mid-19th and through the early 20th centuries. They did not improve by even a notch when the British and the French occupied and then distributed between themselves almost all of the Middle East in 1918-1920, destroying the Arab Kingdom of Syria and crushing all the attempts at independence in Iraq in the process; even less so when they intensified the Jewish settlement in Palestine; crafted Lebanon as a Christian-dominated foothold and granted parts of the French Mandate of Syria to the newly established Republic of Turkey; crushed in blood multiple uprisings in Iraq and Palestine of the 1930s; destroyed the Iraqi armed forces during the Anglo-Iraqi War of 1941, and the least when a disjointed 'alliance' of Arab armed forces was defeated by the newly established nation of Israel, during the First Arab-Israeli War of 1947-1949 (generally known as the 'Palestine War' in the Arab world). Unsurprisingly, most Western narratives, in which the states in question became 'Soviet clients' – indeed 'Soviet puppets' – and thus representatives of the USSR's interests once they began purchasing Soviet arms, are little more than hypocrisy of the finest order. Of course, the establishment of Israel in the de facto centre of the Middle East, and constant Western meddling in local affairs were not the only reasons for successive Arab countries turning to the USSR for armament: political instability and countless coups driven by rivalry between various Arab politicians, or confrontations between Arab states, or between Arab states and their neighbours, have also provided their own impetus. However, there is little doubt that, except in the case of Morocco, defence against Zionist aggression and/or Western imperialism were some of the most-frequently quoted explanations for the acquisition of Soviet arms anywhere from Algiers in the west to Baghdad and Aden in the east, and almost without interruption for at least the last 75 years.

IN SEARCH OF INDEPENDENCE

Nominally at least, it was the government of Egypt in 1955 that made the first step in affairs that led to the emergence of Soviet-made military aircraft in the Middle East. However, explaining the whys by stating something like that in 1955 the Egyptians went to Moscow and bought MiGs and then nationalised the Suez Canal in order to be able to pay for these – and for the construction of the High Dam on the Nile – as frequently explained, would be not only a massive oversimplification, but actually border on a lie.

Conquered by the Ottoman Turks in 1517, Egypt was a province of the Ottoman Empire governed by the Mamluks – a military caste of Turkish and Circassian origin that ruled for centuries – until in 1805 the French expeditionary forces were expelled, and power seized by Muhammad Ali Pasha, an Albanian military commander of the Ottoman Army of Egypt. While officially carrying the title of Viceroy of Egypt, Ali Pasha's subordination to the Ottoman Empire

Naquib and Nasser early after the Egyptian Revolution of 23 July 1952. (Mark Lepko Collection)

was purely nominal: in reality, he massacred the Mamluks (even if leaving their families at peace, and thus enabling them to become a part of 19th century Egyptian elite) and established a dynasty that was to rule the country until 1952. After completely reforming and modernising the economy and military, Ali Pasha – although 'not really' the ruler of Egypt, but a military leader who did his best to reduce the Ottoman influence – annexed northern Sudan and parts of Arabia and Anatolia, only to attract the attention of several European powers: all fearful he might topple the Ottoman Empire without letting them have their own piece of the cake. London and Paris took care that the new ruler of Egypt would return most of his conquests. Subsequently, the government in Cairo – still nominally subjected to the control of the Ottomans – was gradually subverted through patronage and corruption, all supposedly in the interest of the construction of the Suez Canal. Unsurprisingly, by 1875, Ali Pasha's government was bankrupt and could only be saved through the imposition of British and French controllers into the cabinet. In complete ignorance of widespread dissatisfaction with the European intrusion, in 1882 the United Kingdom then invaded Egypt, crushed Ali Pasha's army in the Battle of Tel el-Kebir, and occupied the country: henceforth, the Khedivate of Egypt became a de facto British protectorate under nominal Ottoman sovereignty. Unsurprisingly, considering this ironic status, very little changed for the next four decades: indeed, next to nothing changed even once Egypt – under British pressure – declared its independence in 1914, and the Ottoman Empire collapsed at the end of the First World War, four years later. Certainly enough, Sultan Fuad I assumed the title of King of Egypt, and in 1923 a constitution based on a system of parliamentary monarchy was implemented, but the British maintained their military occupation and retained full control over the government.

British influence was not only the primary reason why the Royal Egyptian Air Force was established only in 1932, as the second official native military flying service in the Arab world (after the Royal Iraqi Air Force, established in 1931), but remained a matter of ever hotter dispute between Cairo and London over the following two decades. Although the mass of British troops were withdrawn to the Suez Canal Zone in 1947, nationalist and anti-British sentiments were further bolstered by the Egyptian defeat in the Palestine War, to a degree where not only the nationalist Wafd Party won the parliamentary elections of 1950, but its government then ordered a withdrawal of all the remaining British troops, and cut off water and supplies for their bases in the Suez Canal Zone. Exploiting the fact that the Egyptian police was observed to be providing arms and support for the local guerrillas involved in attacks on military bases, the British hit back in force on 25 January 1952: their army levelled a police station in Ismailia, killing 43 policemen in the process. The result was an outburst of anti-British riots all over Egypt, in which at least 26 people were killed and over 500 wounded, and over 300 British (or British-related) businesses were torched, a day later. Moreover, although the last monarch of the country, King Farouq subsequently attempted to improve Anglo-Egyptian relations, it was too late: on 23 July 1952, he was overthrown in a military coup staged by a group of Egyptian military men known as the 'Free Officers Movement', in an event ever since known as the 'Egyptian Revolution'.

Motivated foremost by their nationalism, the Free Officers created a nine-member Revolution Command Council (RCC) as a new government, effectively presided over by the leader of their movement, Lieutenant-Colonel Gamal Abdel Nasser. Because Nasser was unknown to the Egyptian public, the RCC initially appointed a hero of the Palestine War, General Mohammed Naquib Yousef Qotp Elkashlan as the President of the newly declared Republic of Egypt, on 26 July 1952. In an attempt to profile himself for the Egyptian public, Nasser then initiated negotiations with London aimed at instrumenting a complete British withdrawal. The talks were concluded between 1953 and 1954 when, in exchange for abandoning its claims to suzerainty over the entire Nile Valley, Cairo received the British agreement to a complete withdrawal of the Suez garrison. When Nasser addressed a mass public gathering in Alexandria to announce the signing of the related agreement on 26 October 1954, he narrowly survived an assassination attempt by the Muslim Brotherhood. Indeed, because of his successful negotiations, and because he maintained his poise amid the resulting chaos and panic, Nasser became the new national hero of Egypt. Emboldened, he not only purged the armed forces of all potential opposition, but had much of the political opposition – including the

Islamists, Leftists and Royalists alike – incarcerated, and the RCC appointed him the new president.

Early on, Nasser's ascent to power in Cairo was not a reason for major concern in the West. The new president was preoccupied with several massive undertakings at home, including a continuous confrontation with the Muslim Brotherhood. However, major differences developed before long. On one hand, Nasser intensified efforts to construct the Nile High Dam near Aswan, as the centrepiece of future economic development of Egypt. On the other, while running secret peace negotiations with the new neighbour to the east of his country, he received reliable intelligence that France was openly providing Israel with ever increasing amounts of armament – in violation of the Tripartite Declaration from 1949, under which Paris, London and Washington agreed to prevent an arms race in the Middle East. While economically unsustainable and surviving only thanks to direct grants-in-aid advanced by the US government, Israel pursued its own designs, the mass of which were expansionist by nature: indeed, repeated raids by the Israeli armed forces on Egyptian military outposts along the armistice lines of the Palestine War, and a campaign of Israeli terror attacks on US and British companies in Egypt (the so-called Lavon Affair) made it clear that there was no serious interest in any negotiated settlement. Meanwhile, although agreeing to withdraw their troops from Egypt, senior British politicians refused to treat Nasser as an equal. Concerned about his growing influence upon British possessions in Africa and Asia and support for nationalist liberation movements, some began considering Egypt under Nasser akin to Nazi Germany under Adolf Hitler in the 1930s. Finally, the USA – still an unknown quantity in the Middle East – monitored the situation in terms of various dogmas, one of which was to insist upon seeing every affair through the prism of the Cold War. Because Nasser was on the pay list of the Central Intelligence Agency (CIA) and incarcerating Egyptian Leftists with gusto, Washington initially expected him to prevent a possible Soviet penetration, and make Egypt a member of a military alliance of Middle Eastern states similar to the North Atlantic Treaty Organisation (NATO) in Europe: this, according to the expectations of the US government, was to 'solve' the Arab-Israeli conflict in a similar fashion to the creation of NATO solving the traditional Franco-German dispute. However, when Nasser turned down all offers of military aid – principally because these were tied to the provision of basing rights for the US armed forces – Washington, in alliance with London, increased pressure upon him by refusing to finance the construction of the Nile High Dam.[2]

Facing growing pressure from several directions, unable to source support from anywhere in the West, and – as a former officer of the Egyptian Army – perfectly aware of an existing qualitative and quantitative superiority of the Israeli Defence Forces (IDF), the President of Egypt was left with no choice but to make a number of fateful decisions within a very short period of time.

EGYPTIAN AIR FORCE OF 1955

Considering these developments, it is unsurprising that the condition of the Egyptian Air Force was gradually degenerating through the early 1950s: its equipment and training remained

Two Fiat G.55 fighters of the EAF: acquired in a great hurry during the Palestine War, they continued serving as advanced trainers until at least 1955. Notably, both aircraft were re-painted in high-speed silver finish overall, and wore yellow 'trainer' stipes around their rear fuselages and wings. The nearest wore the serial number 1208. (David Nicolle Collection)

A brand-new DHC-1 Chipmunk of the EAF seen shortly after delivery in 1953. The type served as the basic trainer until replaced by the Yak-18 in 1958. (David Nicolle Collection)

restricted by what Great Britain was ready to deliver – which, in general, was very little. Out of over 200 military aircraft ordered by Cairo from London between 1949 and 1955, less than a quarter was actually delivered, and these primarily consisted of now-obsolete types. Moreover, stocks of spares and ammunition were critically low, and the service was then seriously damaged by two series of purges: one immediately after the revolution of 1952 resulting in the removal of all officers above the rank of wing commander, and the other in 1953 resulting in the removal of officers of different ranks suspected of sympathising with religious fundamentalists or Leftists. In their place, the RCC began promoting for political loyalty, rather than professional competence. Other changes were of cosmetic nature, and included replacement of the traditional, British-influenced ranking system. Nevertheless, many Air Force officers continued thinking of themselves in what the Egyptians called 'air force style ranks' and is why these remained in widespread use for decades longer.

As of 1952, the shortages were obvious in regards of instructor pilots, training aircraft and tools. Although the Air Force College was moved from Almaza AB outside Cairo to a newly constructed facility at Bilbeis AB, this was short of training manuals, equipment and flight- and ground-instructors. The principal basic trainer was still the Miles Magister, designed and manufactured back in the 1930s. Advanced training was undertaken on survivors of 20 North American T-6 Harvard trainers acquired in the late 1940s, but the British embargoed an order for 15 Canadian-built examples in 1953. The problem with the Magisters was partially solved through the acquisition of 22 de Havilland Canada (DHC) DHC-1 Chipmunk basic trainers in 1953, and then Factory 72 in Heliopolis acquired a manufacturing licence for the German Bücker Bü.181D Bestman basic trainer in the form of the Gomhouriya (Republic) T.Mk 1. Nevertheless, remaining Texans had to be kept in service for a while longer, and the Egyptians were forced to press into service some older piston-engined fighters – including worn out Fiat G.55s and Aer Macchi (subsequently Aermacchi) MC.205s, Supermarine Spitfire F.Mk 22s, and Hawker Sea Furies – for advanced training. While many Egyptian officers were keen to maintain contacts with their counterparts in the Royal Air Force (RAF) units still deployed in the Suez Canal Zone, the British were anything but enthusiastic, and only sporadically provided a small number of instructors. As a consequence, in 1954 Cairo reached an agreement with New Delhi, under which groups of instructor pilots of the Indian Air Force (IAF) began serving as instructors for basic flight training at the Air Force College in Bilbeis.

The fighter fleet of the EAF was in a similar condition: at the time of the revolution in 1952, the premier fighter jets were the survivors of 12 Gloster Meteor F.Mk 4s and 6 T.Mk 7 two-seat conversion trainers acquired in 1949-1950, which served with No. 20 Squadron. In late 1952, London granted permission for delivery of 12 refurbished, ex-RAF Meteor F.Mk 8s: only four were delivered before a temporary arms embargo, and it was only in 1955 that the remaining eight aircraft reached Egypt. The backbone of the Air Force was thus formed by de Havilland Vampire FB.Mk 52s, with 20 out of 50 that were on order from London delivered starting in 1950, before the others – and an agreement for their licence manufacture in Egypt – were all cancelled by another embargo. Left without choice, and following lengthy negotiations, the Egyptians then acquired 58 refurbished examples from Italy – under the guise of deliveries to Syria. As of 1955, the EAF thus had two operational squadrons equipped with Vampires: Nos. 2 and 31. Furthermore, the Fighter Training Unit (FTU) at Almaza AB was equipped with a miscellany of Vampire FB.Mk 5s, FB.Mk 52s and T.Mk 55s. Because early Egyptian attempts to acquire night-fighters had been blocked by the British arms embargo, it was only in 1955 that London finally agreed to sell six Meteor NF.Mk 13s. These arrived in early 1956, but only two are known to have entered service, and the unit operating them – No. 31 Squadron – had only one pilot that underwent a night-fighter Course in the UK. The night fighting capability of the Air Force thus remained very limited, and there was not enough equipment and spares to maintain the complex AI.10 radar set installed in the stretched nose of the NF.Mk 13.

One of 56 Italian-manufactured Vampire FB.Mk 52s (serial number 1529), operated by Nos. 2 and 31 Squadrons, seen in flight after delivery to Egypt. Italian-made Vampires in Egypt were easily distinguished from the British-made examples by their prominent black anti-glare panel applied in front of the cockpit. (Albert Grandolini Collection)

BOMBER AND TRANSPORT FLEETS

Although permitting the export of nine Avro Lancaster B.Mk 1s and nine Handley Page Halifax A.Mk 1s to Egypt in the late 1940s, Britain subsequently adapted a policy of consistently opposing Egyptian attempts to obtain an effective bomber capability. Indeed, the British proved extremely upset when the REAF added guns to the turrets of aircraft delivered as unarmed, and officially sold for meteorological- and air-sea rescue duties. While not considering them a threat, the British knew that, in Egyptian eyes these aircraft formed a training nucleus for a future, more effective bomber force. Therefore, they refused to deliver even spare parts for these aircraft and most had to be grounded by 1953. While still consisting of at least three Lancasters and three Halifaxes that were kept in 'flyable' condition, the EAF's bomber fleet had de facto ceased to exist by 1955.

Despite efforts to standardise its inventory, the EAF still operated a remarkable mix of about 40 transport and support aircraft. These included about 20 Curtiss C-46 Commandos and a similar number of Douglas C-47 Dakotas operated by combined Nos. 7 and 12 (Transport & Paratroop) Squadrons, and No. 3 Squadron (Transport), respectively, and their crews were trained in close cooperation with the first Egyptian paratrooper units, established in the early 1950s.[3] These were complemented by Beech C-45 light transports operated by No. 4 Squadron, but principally used by the Navigational School set up at Dikhelia AB, outside Alexandria.

The luxuriously equipped ex-Royal Flight Dakota had been given the new serial number 113, and entered service with what was

A bottom view of a trio of Vampire FB.Mk 52 fighter-bombers of the EAF. Notable are not only the identification stripes around the wingtips but also the launch rails for unguided rockets installed behind the intakes: these were a genuine Egyptian modification. (Albert Grandolini Collection)

now No. 11 (Communication) Squadron, along with two ex-Royal Flight Westland-Sikorsky S.Mk 51B helicopters. Although only one of the latter was still operational, and there were only two officers qualified to fly them, the S.Mk 51 was based at Almaza and served for search and rescue purposes as of 1955. Other aircraft of the former Royal Flight – including Grumman Mallards, four Dakotas and one Commando – served with the same unit.

DIVIDE AND RULE OF SYRIA

While not as famous for major architectural achievements, the civilisation in the area known as Sham in the Arab world, and Syria since the times of the Roman Empire, is at least as old as that of Sumeria, Akkadia, Assyria, Babylonia, ancient Egypt and the Hittites. Due to its geographic location in between such powers, Syria became broadly synonymous with the Levant: a centrepiece in the traditional trade, which resulted in the emergence of thriving urban centres like Aleppo, Damascus and Palmyra – together with their natural ports of Ugarit (replaced by Latakia), Tripoli and Beirut. Unsurprisingly, it was also the scene of countless invasions until becoming the centre of the Arab world during the times of the Umayyad Caliphate, in the 7th century. Savaged by the Crusades and then Mongol invasions in the 11th and 12th centuries, and now populated by a colourful mix of ethnic and religious groups, the area came under Ottoman control in the 16th century. Over the following 400 years, it gradually recovered into one of the Ottomans' most prosperous provinces. Rigorous economic, educational, and political reforms of the Ottoman Empire in the late 19th century resulted in a flourishing economy, and the emergence of a meritocracy, which established many of the locals in distinguished positions in the civilian and military administration – and the revival of Arab nationalism. Nevertheless, thousands of officers and other ranks of Syrian origin served in the Ottoman Army during the First World War and fought with determination and distinction against British and Allied forces in Gallipoli and Palestine.

Although their forces failed to make any advances during the first two years of the First World War, in 1916 Great Britain, France and Russia secretly agreed on the post-war division of the Ottoman Empire in the form of the Sykes-Picot Treaty, in which – and amongst other points – France was to gain possession over all of the eastern Mediterranean coast, most of Anatolia and northern Iraq, while the British were to establish themselves in control over nearly everything south of that area, from Palestine in the west to the Persian border in the east. The Russian Empire collapsed in 1917, but near the end of the First World War, British-supported forces of the Sharifian Army led by Emir Faysal Ibn Hussein, son of Sharif Hussein Ibn Ali of Mecca, entered Damascus. Contrary to Western designs, Faysal then established the Arab Kingdom of Syria in 1919-1920, with the support of local nationalists, as a constitutional monarchy encompassing all of the former Ottoman possessions from Antioch in the north to Aqaba in the south. However, insistent on their own agreements and designs, London and Paris persisted: when the French landed their invasion forces in Beirut, the British left the Arab Kingdom of Syria and their ally Faysal to their own devices.

Contrary to the Empire, the Ottoman Army did not collapse at the end of the First World War: in 1919, it rallied around General Mustafa Kemal (best-known as Atatürk, 'Father of Turks') and offered effective resistance to the French invasion of Asia Minor. Emboldened, Atatürk's former comrades-in-arms further south – foremost General Yusuf al-Azma, of Damascene origins and now serving as the Minister of War and Chief of Staff of the Army of the Arab Kingdom of Syria – were keen to repeat the feat in Syria. However, Faysal folded, causing disorder in his armed forces. What was left of the latter was then defeated in the Battle of Maysalun, the decisive clash with the French forces advancing from Beirut on Damascus, fought on 24 July 1920.

Although in firm control of Damascus only days later, the French promptly found themselves facing major armed uprisings almost everywhere from Alexandretta and Antioch (present day Iskenderun and Antakya in Turkey, respectively), via Aleppo and the mountains above Latakia, to Hama, Homs, Damascus and to Dayr az-Zawr, in 1920-1925. While suppressing these through terrorising the local population and the widespread destruction of private possessions, they were forced into negotiations with leading nationalists – even if for no other purpose than to buy time. Meanwhile, seeking to secure support and prolong their presence, the French sought to cause a rift between the predominantly Sunni Muslim population and different ethnic and religious minorities: the latter were granted their own statelets – primarily Lebanon, supposed to serve as a Christian stronghold although predominantly populated by Muslims. Along the same lines, the French began superimposing the 'value' of (supposed) 'war-like races' – foremost the Alawis and the Druze – and drafting these for the establishment of auxiliary forces serving the purpose of safeguarding domestic security. In 1933 they established the Military Academy of Homs for the training of the necessary officers and non-commissioned officers (NCOs).

Syria – and Lebanon – thus remained within a tight French grip for decades longer: even the Treaty of Independence from September 1936 was never ratified by Paris. It was only a combination of continuous political unrest, frequent strikes and riots, and severe pressure from London that prompted the French authorities to let Syrians proclaim their independence in the form of a parliamentary republic in 1941, and then finally withdraw their last troops from the country in 1946. The young Syrian democracy was quickly destabilised by the defeat of its armed forces and the loss of possessions in the Palestine War, and the activities of the CIA of the USA, resulting in a series of coups, including two in 1949, one in 1951 and another in 1954. Ultimately, successive governments became overdependent on the loyalty of the armed forces for their own survival to the point where they had little option but to secure this through significant arms acquisitions abroad.

SYRIAN ARAB AIR FORCE OF 1955

As early as 1943, leading Arab nationalists of Syria developed a plan for establishing an army of three divisions that would be deployed to support the Allied forces against the Axis forces in Europe. The French took great care to prevent such designs and thus upon independence Damascus inherited only about 8,000 officers and other ranks of the auxiliary force named the *Troupes Speciales*. At that point in time, and in a country populated up to 80 percent by Sunni Muslims, Alawis dominated the nascent Syrian armed forces. Their dominance of the Syrian Arab Army was to further increase during the 1950s and the 1960s, eventually reaching the point where Alawi officers completely excluded the traditional Sunni leadership from top positions in the military hierarchy – and thus from the mainstream of Syrian politics.[4]

Another factor exercising strong influence upon the development of the Syrian armed forces was the appearance of several German advisors – mostly veterans of the *Abwehr*, *Gestapo*, *Schützstaffel* (SS) and *Waffen-SS*, and the *Wehrmacht* during the Second World War – de facto imported to the country by the CIA in the late 1940s with the aim of exercising influence over the local political scene. Although most of them left by the mid-1950s, they laid the fundaments of the police state as known until this very day.[5]

German advisors also played an important role in the build-up of the SyAAF. Officially established on 16 October 1946, this service grew very slowly at first, while supervised by a small group of British and French advisors – until its first commander, Colonel Abdel Wahhad al-Hakim, contracted two German and five Croat pilots found in one of the many camps for displaced persons in Italy in 1947. Led by Fritz Strehle (an 'ace' of the Second World War and a former Messerschmitt Me.262 pilot) and Mato Dukovac (top Croat ace of the Second World War) these trained the first group of 30 Syrian pilots to fly a miscellany of Piper J-3C-64 Cubs, Percival Proctor Mk Vs and Tiger Moths acquired from the USA and Great Britain. Operating from Estabel airfield in Lebanon during the Palestine War, Dukovac and his students then flew combat sorties in 20 North American T-6 Texans meanwhile bought from the USA – sometimes in cooperation with Avro Anson light transports of the Royal Iraqi Air Force, forward-deployed at the al-Mezzeh airport outside Damascus.[6]

The first aircraft to ever wear Syrian national insignia were four ex-RAF Percival Proctor Mk IVs purchased in 1946. The insignia had the form of the national crest used until 1958, and the aircraft wore the registration SR-A1 (shown here), SR-A2, SR-A3 and SR-A4. (Albert Grandolini Collection)

In 1950, Syria acquired 30 Fiat G.55 and Fiat G.59 fighters from Italy, followed by G.59/2B two-seat conversion trainers. The G.55s and G.59s mainly served with No. 1 Squadron, then commanded by Squadron Leader Kumani, while the G.59/2Bs were assigned as advanced trainers to the Air Force College. (David Nicolle Collection)

Following the Palestine War and two military coups d'état of 1949, the SyAAF experienced a period of at least relative growth. In 1950, after acquiring 26 Fiat G.55s and four Fiat G.59B-2 fighters from Italy, followed by nine Fiat G.46-1 two-seat conversion trainers, and helping Egypt secure the above-mentioned acquisition of Vampire FB.Mk 52s, Damascus signed a contract with London providing for training of pilots and ground personnel in Great Britain, and the acquisition of about a dozen de Havilland Chipmunk basic trainers, 10 reconditioned Spitfire F.Mk 22 fighters, 12 Meteor F.Mk 8 jet fighters, and two Meteor T.Mk 7 conversion trainers. However, before long the British government began treating Syria in similar fashion to Egypt and subjected it to a

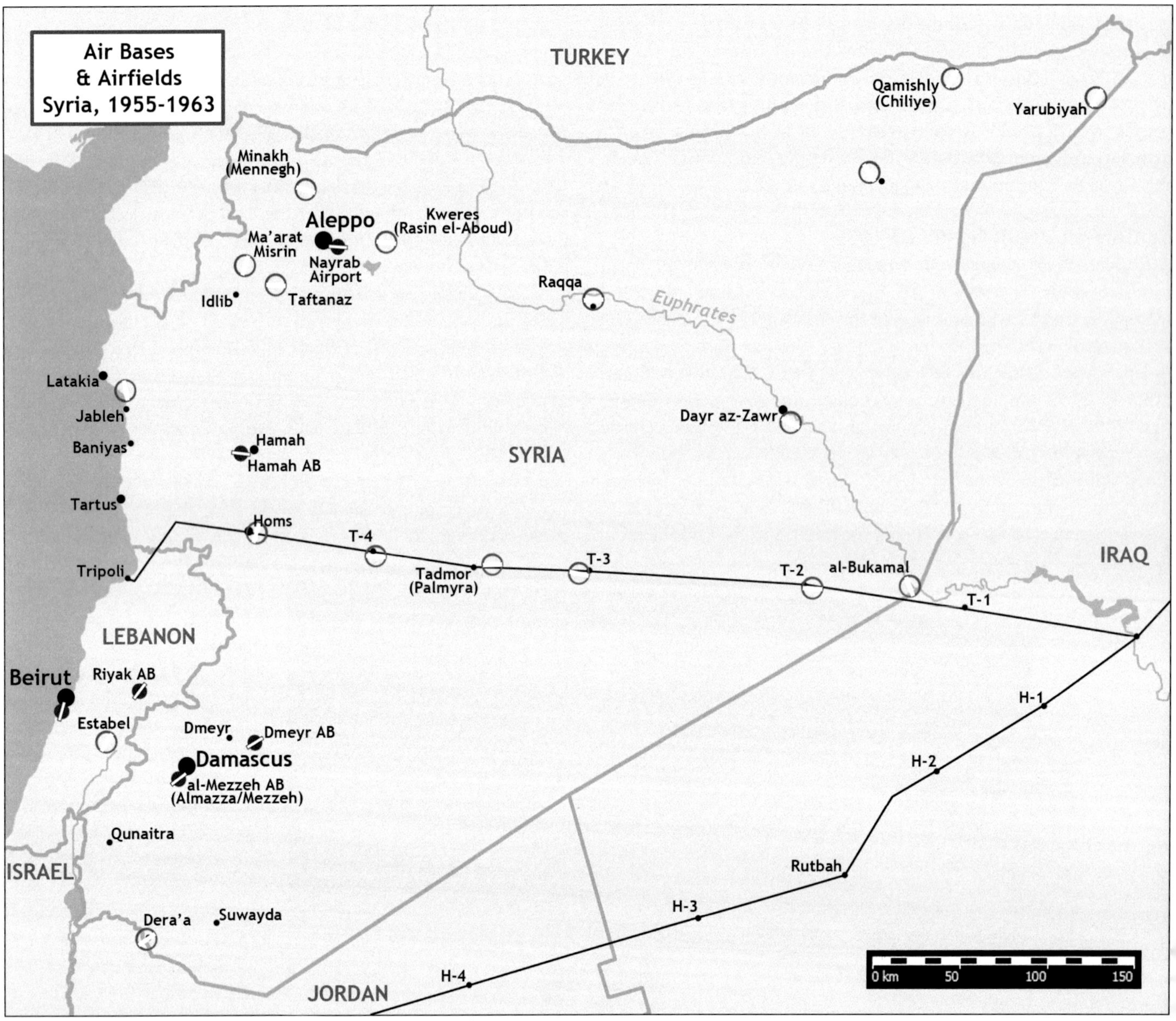

A map of air bases and other airfields in Syria in 1955-1963. (Map by Tom Cooper)

series of on-off embargoes. While Chipmunks and Spitfires were delivered, Meteors were embargoed until 1952, when Damascus rushed to buy additional ones. By 1953, a total of 18 such jet fighters were in Syria: although the British then withdrew their advisors and all the night-fighters arrived without their radars, the unit operating them – No. 9 Squadron – was worked up with some support from Egypt. A year later, the Air Force College was established at the military side of Nayrab Airport, outside Aleppo and its first class included 15 graduates of the Homs Military Academy. That said, and contrary to the EAF, the SyAAF had not a single operational radar system in service, and bar three, most of the available air bases had runways too short for jet operations.

RAPIDLY GOING NOWHERE

Overall, as of 1955 the first two air forces in the Middle East to acquire Soviet-designed combat aircraft were almost as colourful as the fleets they flew: they came into being at the insistence of local officers, and although almost exclusively equipped with aircraft of British and US origin, they grew *despite*, rather than *with* the help of British and French advice and instruction. Still, and ironically, both the EAF and the SyAAF had their traditions and operational practices strongly influenced by the British in particular – and this went well beyond their organisational structures, ranks and insignia, uniforms, even their breakfasts. Due to the fact that both the British in Egypt and the French in Syria did their utmost to completely delete any kind of recollections of historic native military leaders and traditions – especially officers serving with the Ottoman Army in the early 20th century – there were next to no local military theoreticians. The senior leadership of both services was trained abroad, foremost in Great Britain, and using British military literature for guidance. The result was a way of thinking heavily orientating upon what the Westerners taught them: the foremost was that Arabs could never do anything right and could never establish an air force comparable to the RAF, for example. Even once they were free of British and French meddling, repeated purges of the top ranks of both services stripped them of skilled administrators, the few influential strategists and tacticians that were around, and experienced disciplinarians and instructors, effectively de-professionalising them in the process. Those that were left to continue serving were only of a calibre sufficient to oversimplify and conclude that if they would equip and train the way the RAF was equipped and trained, they 'must' become as good as the RAF. Following this line of thinking, and also because of the lack of influence of the EAF officers in Cairo, the widespread preoccupation of the Alawite officer-cadre with politics

HOSNI MUBARAK: A FATHER FIGURE – OR A VILLAIN?

Nowadays primarily remembered for his political career, the military career of Hosni Mubarak, President of Egypt from 1981 until 2011, remains next to unknown to the general public. In Egypt, he is remembered as the victor of the October 1973 Arab-Israeli War, or a dictator with much blood on his hands. In the West, very few recall him as commander of the Egyptian Air Force from 1972 until 1975.

Born on 4 May 1928 in Kfar el-Meselha, Mounufia Governorate in the Nile Delta, Hosni Mubarak graduated from the Military Academy in 1949, joined the Air Force College in 1949, and graduated on 13 March 1950. The official story in Egypt is that Mubarak subsequently spent two years in a Spitfire fighter squadron. Certainly enough, the Royal Egyptian Air Force operated various marks of Spitfire since the mid-1940s, but most of these were already worn out by the time of the Palestine War. Mubarak actually served as instructor pilot with the Fighter Conversion Unit (FCU) at Almaza AB (nowadays a suburb of Cairo and the site of the EAF Museum), and then taught elementary and advanced training to no fewer than seven generations of future pilots. It was in this function that most older EAF pilots remember him – and it was a crucial position at the time the Egyptian Air Force was critically short on experienced instructor pilots.

By 1955, Mubarak was assigned to the Staff of the Air Force College, where he served during the Suez War of 1956. A year later, he was re-qualified to multi-engine and jet-powered aircraft, which resulted in his 'ultimate specialisation': after helping work up the Air Force Academy, in 1959 he was sent for conversion training to Lugovaya AB in the USSR for a conversion course on Ilyushin Il-28 bombers. Upon his return, he was assigned the command of the Abu Suweir-based Air Group 61, including three squadrons equipped with Il-28s. By 1961, Mubarak was back at Ryazan AB in the USSR, together with a group of 10 other Egyptian crews, to undergo a four month conversion course to Tupolev Tu-16 bombers. Back in Egypt as of 1962, he resumed the command of Air Group 61 in time to lead it during the early phase of the Yemen Civil War. Before long, the unit was reinforced by Soviet-owned and operated Tu-16s, and Mubarak began flying joint operations. In March 1964, Hossni Mubarak entered a staff course at the Frunze Military Academy in the USSR: following graduation in 1965, he returned to Egypt to work-up the first unit equipped with 24 Tu-16s: Air Group 65, which flew several nocturnal strikes against targets in Saudi Arabia. As a primary strike force of the Egyptian Air Force, Air Group 65 was the primary target of the Israeli Air Force during the June 1967 War, and had 17 of its bombers destroyed in the first minutes of the war, all on the ground. By accident, the other six survived only because minutes before the attack a formation led by Mubarak launched from Cairo West for a training sortie over southern Egypt. He led his crews into a successful landing at Luxor Airport, undertaken without any kind of ground control, but then made a crucial mistake in making a telephone call to Cairo to report to the High Command. Because the Israelis had tapped all the important telephone cables in Egypt, they thus received first-hand intelligence about the presence of six Tu-16s at Luxor. Less than one hour later, their fighter-bombers swooped in to destroy all the bombers on the ground.

After the June 1967 War, Mubarak and his Tu-16 crews were all sent to Iraq for continuation training on bombers of the Iraqi Air Force. In this fashion, he avoided several purges of the top ranks of the Egyptian armed forces that took place during the following months. Indeed, renowned as a good administrator, excellent instructor and a professional military officer, he was returned to Egypt in 1969 and then not only assigned the responsibility of rebuilding the Tu-16 fleet but also appointed the Chief of Staff in June 1969, and then the commander of the Air Force in April 1972. It was in this function that Hosni Mubarak led the service during the October 1973 War with Israel, and until his retirement from active service and appointment as Vice-President of Egypt in November of the same year.

Although of poor quality, this rare photograph shows a group of flight instructors at the Air Force College in Bilbeis around 1955, in front of a Spitfire F.Mk 22 (left, background), and Fury FB.Mk 11 (right, background). Hosni Mubarak stands in the rear row, second from right. (via Dr Abdallah Emran)

in Damascus and the haphazard way in which both services were equipped and trained, next to nobody paid attention to modern theories of aerial warfare, and even less so spent time thinking about how to defeat their major common enemy – the, at least in theory, equally nascent IDF/AF. Strongly influenced by stories about the Battle of Britain, Egyptian and Syrian officers expected that the EAF and the SyAAF would thus have enough opportunity to pick the time and place of the fight.

2

CZECHOSLOVAK ARMS DEALS

Hampered by Western arms embargoes, during the early 1950s both the Egyptians and Syrians launched several attempts to acquire arms from the countries of Eastern Europe.[1] However, Soviet leader Joseph Vissarionovich Stalin did not want to become involved in the Middle East for fear of British retaliation. Furthermore, the Soviet officials involved considered the Arab requests as attempts to play the Soviets off against the British. Therefore, first the Syrians and then the Egyptians attempted to negotiate with Czechoslovakia – the same country that was renowned as the major arms supplier to Israel in the late 1940s. Initially at least, Prague turned down all requests: Czechoslovakia was now under the firm control of the USSR. The situation began to change only with the rise to power of Stalin's successor, Nikita Sergeyevich Khrushchev, in 1953. Khrushchev not only abandoned the old policy of considering all non-communist countries as enemies but decided to play a more active role in the Middle East and encouraged the Czechoslovak government to strengthen its ties with Egypt and Syria. Even then, successive visits by Egyptian delegations proved fruitless: it was only in August 1954 that the Soviets signalled their preparedness for serious negotiations and even then, Nasser was discouraged to a degree where, by early 1955, he opened secret negotiations with Great Britain and the USA, requesting not only arms but above all help with financing the High Dam at Aswan. By then he was behind the Syrians in the queue for Czech arms.

OPERATION 104

With hindsight and considering the kind of shockwaves that the famous arms deal between Cairo and Prague of 1955 sent all over the world, it is certain to surprise that it was Syria, not Egypt, that 'opened the doors' to acquiring armament from East European nations, and thus also the USSR. Still influenced by its German advisors, in 1953 the Syrian armed forces acquired a small number of *Panzerkampfwagen* IV (*PzKpfw* IV, colloquially 'Panzer IV' in English) tanks and Hummel self-propelled guns from France. The vehicles arrived in poor technical condition and were soon in need of replacement: indeed, none was left in operational service by the time of the February 1954 military coup against the pan-Arabist government of Colonel Adib ash-Shishakli, plotted by members of the Syrian Communist Party, Druze officers and members of the Ba'ath Party. In an attempt to satisfy the demands of the armed forces and the disparate political parties supporting him, the new President – Hashim Qalid al-Atassi – sought for ways to improve relations with Eastern Europe. The first to react was Prague: in mid-1954, Czechoslovakia re-established its embassy in Damascus and over the following months entered talks with representatives of the Syrian armed forces, who inquired about possible deliveries of German-made tanks left in Czechoslovak stocks from the times of the Second World War. Related negotiations were concluded in May 1955, when Damascus placed an order for 45 Panzer IVs and 12 *Sturmgeschütz* III assault guns, with associated spares. No sooner were the Panzers delivered – by ship from the Soviet port of Nikolayev in the Black Sea – the Syrians began demanding more, and then issued their first request for the delivery of Soviet-designed armament. Initially, the Czechoslovaks were reluctant and attempted to satisfy the Syrians with deliveries of additional German-made tanks, small arms, and spare parts: only months later – apparently before the first similar deal was signed with Egypt – Damascus and Prague signed a contract for the delivery of, amongst others, 41 MiG-15bis and four MiG-15UTIs. Thus began what the Czechoslovak government designated 'Operation 104': its military cooperation with Syria.

ISRAELI RAID ON GAZA

While the Syrian arms deal with Czechoslovakia completely evaded public attention in the West, the situation in Egypt began resembling that inside a kettle filled with boiling water. In reaction to Nasser's secret requests for arms, in early 1955 the USA and Great Britain offered to finance construction of the High Dam, and the USA also expressed a preparedness to sell some arms – but only on condition of Egypt providing them with basing rights and joining an anti-Soviet alliance. Unsurprisingly, Nasser countered with requests of his own, most of which proved unacceptable for the West. Negotiations dragged on without the parties getting any nearer a common accord. In an attempt to find a solution, British Prime Minister Anthony Eden – an Arabic speaker who considered himself sympathetic towards the Arabs, had never supported Zionism and who had welcomed the expulsion of King Farouq – visited Egypt in January 1955. However, on arrival in Cairo, he treated Nasser in a patronising fashion. Unsurprisingly, Eden's greeting in flawless Arabic left the Egyptian unimpressed, and prompted him to express his hostility towards the West attempting to draw Arabs into the Cold War.

Then, late in the evening of 27 February 1955, the IDF raided the Gaza Strip, killing 46 Egyptian soldiers and wounding 29, most of them still in their beds. What became colloquially known as the 'Gaza Raid' proved to be a turning point for Nasser: it not only forced him into a realisation that a negotiated settlement of the Arab-Israeli dispute was impossible but caused him to expect a new war with Israel: however, for this purpose his humiliated and now enraged military was clearly in need of new arms. Indeed, this became his priority number one. Only a few days later, and under conditions of utmost secrecy, emissaries of the RCC met Czechoslovak and Soviet representatives in Cairo to discuss a possible arms deal. Even then, and even after the visit to Prague by the Deputy Minister of War – Hassan Ragab – on 6 April 1955 and a positive reply from Moscow six days later, negotiations were protracted. For all practical purposes, the Egyptians were still reluctant: indeed, suspecting that once their armed forces were reliant on the Soviet arms, Moscow might misuse Egypt's need for ammunition and spares to exercise political influence, Nasser continued secret negotiations with the USA throughout. However, when it became obvious that the Americans were not ready to abandon their demands for basing rights in Egypt, and the Central Committee of the Communist Party of the USSR authorised the sale of Mikoyan i Gurevich MiG-15 fighter jets to Egypt in mid-June 1955, there was no reason to hold back. On 20 August 1955, an Egyptian delegation travelled to Prague to finalise remaining issues and sign the contract, stipulating deliveries of fighter jets, bombers, guns, mortars, tanks, armoured personnel carriers and associated spares and ammunition between October 1955 and January 1956. Thus came into being the so-called

'Czech' or 'Czechoslovak Arms Deal' that was run as 'Operation 105' in Prague: an agreement for Egypt to acquire Soviet-designed armament from Czechoslovak and Soviet production.[2]

Table 1: List of Aircraft ordered for the EAF in the First Czechoslovak Arms Deal, 12 September 1955[3]

Aircraft or Weapon Type	Number
MiG-15bis*	80
MiG-15UTI*	6
Il-28	45
Il-28U	4
Yak-11	25
Il-14	20

Clandestinely taken by the British at Abu Suweir AB, this photograph is of one of 80 MiG-15bis ordered by and delivered to Egypt in an arms deal that caused uproar in the West. Notably, the jet has received large roundels and identification stripes on the rear fuselage. (David Nicolle Collection)

OPERATION 105

While several small groups of Egyptian instructor pilots underwent hurried conversion courses for new aircraft in Czechoslovakia, most conversion training was to be provided by Czechoslovak and a handful of Soviet advisors in Egypt. The number of MiG-15s and Il-28s that the EAF ordered seems to have been based on a plan for the Air Force to establish three squadrons equipped with jet fighters and three equipped with jet bombers. Each of the planned MiG squadrons was to receive 25 aircraft, of which 12 would be in service, four held in operational reserve, four in secondary reserve and five in storage. Similarly, each of the planned Il-28 squadrons was to have 12 operational aircraft, and three in reserve, undergoing maintenance or in storage.

The first batch of MiGs arrived in Alexandria on 1 October 1955, on board the Soviet freighter *Stalingrad*. They were then taken to Dikhelia AB by road and stored pending the arrival of a small Czechoslovak team (including nine technicians) tasked with assembling them. As the home of the EAF's Navigational Training Base, Dikhelia was protected by one or two flights of Meteor F.Mk 4s from No. 20 Squadron during this sensitive period. As a result, elements of this unit became the first to convert to MiGs, and then served as the Fighter or Operational Conversion Unit (OCU) before being re-organised as No. 1 Squadron.

All 86 MiG-15s were in Egypt, assembled and test-flown by the end of 1955 and, contrary to what was expected by most of foreign observers, the first group of Egyptian pilots adjusted relatively well to the new equipment. The first two squadrons equipped with MiG-15s – Nos. 1 and 30 – were established at Almaza AB in December 1955. On 15 January 1956, eight MiGs made their first public appearance in the form of a short overflight over Cairo.[4] The next day, the same group of Egyptian pilots flew their eight mounts over Bilbeis AB in a formation representing the Egyptian flag for the benefit of President Nasser. On the last day of the month, the Czechoslovak Ambassador to Cairo presented the President with a five-seat Mraz Sokol, while the Soviet Union provided a brand-new Il-14 VIP-transport – serial number 1101 (the same serial number as had been given to one of the earlier C-46 transports) – which replaced the weary Dakota, serial number 113, previously used for such purposes.

Commander of No. 1 Squadron, EAF as of 1955-1956, Shalaby el-Hinnawy, with one of the Czechoslovak or Soviet instructors, in front of two brand-new MiG-15bis. Notable is the lack of serial number on the nose of the aircraft directly behind Hinnawy, and – on the aircraft in the background – the large roundel applied on these early Egyptian MiGs: aircraft delivered after 1956 received much smaller roundels. (Shalaby el-Hinnawy, via David Nicolle)

NASSER'S INDISCRETION AND FURTHER ARMS DEALS

If the initial conversion of two EAF squadrons to MiG-15s progressed at a relatively good pace, this was not to be the case with what was to follow – despite the arrival of four teams of Czechoslovak advisors and at least one of Soviets, starting in late February 1956. The reasons were manifold and included not only the general state of chaos in the Egyptian armed forces, but also an entire series of additional arms deals agreed between Cairo, Moscow and Prague and the behaviour of the Egyptian President Nasser. On 27 September 1955, Nasser – against agreement with Moscow and Prague, though at the end of his patience with Western emissaries, and in the light of reports about an increased flow of French arms to Israel – announced the Czech Arms Deal in public. While including relatively modern combat aircraft, the majority of the armament ordered by Egypt was actually old though abundant, relatively simple and cheap. Nevertheless, Nasser's announcement of the Czech Arms Deal – while prompting an outburst of support at home – sent shockwaves around the world. While understanding that the absorptive capacity of the Egyptian military was limited in the short term, and that it would take several years for Egypt to become capable of making full use of Soviet arms, even professional circles in London and Washington were concerned to the point where both governments sent a number of high-ranking officials to Cairo in a vain attempt to convince Nasser to withdraw his decision: at the time, not only racism, but especially the conviction was widespread that any kind of cooperation with the USSR meant slipping into the Soviet sphere of influence and accepting the communist ideology. In this regard, Western powers expected nothing less from Egypt. Curiously, statements by a handful of moderate US officials were entirely ignored, some of whom reacted with expectations that this deal might signal the start of a new era in the Middle East. Secretary of State Dulles went as far as to observe that Israel would henceforth, 'have to play the part of the good neighbour to the Arabs and not seek to maintain itself by its own force and foreign backing.'[5] Certainly enough, such thoughts were an absolute anathema for David Ben-Gurion, the first Prime Minister of Israel, and its most dominant political figure during the 1950s. He reacted by advising the Chief of Staff IDF to prepare for a war against Egypt, to be fought in summer 1956.[6]

While this remained unknown in Cairo, Nasser continued ordering arms from the new source. In November 1955, the RCC concluded a deal with Moscow for the acquisition of 16 warships and three submarines. In January 1956, this was followed by Cairo and Prague signing their second contract, this time stipulating deliveries of additional ammunition and spares necessary to bolster Egypt's meagre reserve stocks, and in April a supplementary contract was signed in Prague for five Ilyushin Il-14 transport aircraft, together with a sizeable shipment of ammunition for the MiG-15s and Il-28s. In May 1956, the second contract between Egypt and the Soviet Union was signed that – amongst other items – envisaged deliveries of 24 more powerful MiG-17Fs and 26 Czechoslovak-made Mraz Sokol basic trainers. By August of the same year, two further orders were placed – both seeing only a formal Czechoslovak participation – stipulating deliveries of additional spares and ammunition, and the establishment of licence production of small arms ammunition and artillery ammunition in Egypt.[7]

No fewer than 12 MiG-15bis can be seen in this photograph taken during the annual parade celebrating the Day of Revolution, on 26 July 1956. (Albert Grandolini Collection)

One of the brand-new Yakovlev Yak-11s (serial number 529), seen during a ceremony marking their introduction to service in 1956. In Egyptian service, all the Yak-11s were painted in light blue overall and wore yellow 'trainer stripes'. (Nour Bardai Collection)

A long row of Sokol Mraz basic trainers of the EAF, seen shortly after their delivery. Notable are the serials apparently running from 1 up to 25 or 26, and a miscellany of Doves and C-47 Dakotas around them. (David Nicolle Collection)

INCIDENTS

As Israel initiated its preparations for a major war with Egypt, the IDF increased the tempo of provocations over the Egyptian-Israeli armistice lines – always under the pretence of retaliating against Arab state-sponsored terrorism. Certainly enough, and at least officially, no Arab statesman of the time could afford to declare himself as anything less than a staunch enemy of Israel, however, none ever ordered nor sanctioned any kind of terrorist attacks. On the contrary: in great secrecy, both Jordan and Egypt were running peace negotiations and reports by United Nations (UN) and US observers repeatedly stressed that all the governments and armed forces of countries neighbouring Israel were doing their utmost to prevent anybody from crossing the armistice lines. However, knowing well about Israel's military superiority, Ben-Gurion was keen to not only spoil the negotiations, but also to provoke a war that would enable a further expansion of Israel – into Lebanon, into Syria, into Jordan and into Egypt. Correspondingly, he encouraged IDF commanders into provoking the Egyptian armed forces.

On 21 July 1955, during UN-sponsored talks between Egyptian and Israeli military officials, four Israeli aircraft flew over Gaza City. As one of them flew lower than the rest of the formation, it was engaged by Egyptian anti-aircraft artillery (AAA) but was not hit. On 29 August 1955, two Israeli jets overflew Rafah, and were again engaged by Egyptian AAA. Regarding such Israeli provocations as obvious violations of the ceasefire and as efforts to sabotage any kind of negotiations with Egypt, for the first time since the end of the Palestine War the General Command in Cairo granted permission for the EAF to be ordered into action should the need arise, and began bolstering Egyptian armed forces deployed in the Gaza Strip. It expanded the Palestinian Border Guard and established a Fedayeen of about 100 men with the intention of carrying out reconnaissance inside Israel. Unimpressed, on 22 August 1955 the Israelis provoked another clash, during which they captured 25 Egyptian soldiers. The EAF deployed two Vampires from el-Arish to reconnoitre the area on 29 August but these were promptly ambushed by a pair of Israeli Meteor F.Mk 8s and one was shot down close to the ceasefire line, killing its pilot. With skies clear of Egyptian interceptors, the IDF then assaulted Khan Yunis police station, south of Gaza, on 30 August 1955, killing 39 Egyptian troops and civilians, and injuring 16 others. No. 31 Squadron at el-Arish was ordered to retaliate by attacking Kibutz Karmiyya with four Vampires on 1 September. After completing their attack runs, two of these were intercepted by IDF/AF Meteors while flying southward at an altitude of 7,000 metres. One – serial number 1569 – was shot down by cannon fire on the first pass and fell to the ground near Karmiyya. The other – serial number 1567 – pulled hard to avoid the enemy fire before diving in order to accelerate and run away. However, the Egyptian aircraft was much slower than the Israeli Meteors which quickly reduced the range to 200 metres. As the Vampire turned again, this time in an eastern direction, it was hit in the cockpit and came down near the neighbouring Kibbutz Ziqim. Both EAF pilots were killed.

Israeli troops detaining a 'terrorist': one of thousands of Arab civilians that attempted to cross the armistice line either to return to his home, or to work on his field in what once used to be Palestine and was now Israel. Although next to none of the people involved in such illegal crossings of the ceasefire line was armed or had ever attacked any Israeli, the local government registered every single such act as a 'terrorist attack'. (Mark Lepko Collection)

On 28 September 1955, only a day after the announcement of the Czech Arms Deal and in clear violation of the 1949 Armistice

A group of EAF instructors and students seen in front of a row of Yak-11 trainers: the type proved a handful to fly and caused a number of fatal incidents during its early service in Egypt. (Nour Bardai Collection)

Agreement, the IDF moved its troops into the de-militarised zone (DMZ) of el-Auja. When the UN observers eventually managed to convince the Israelis to withdraw, the IDF assaulted the Egyptian outpost at Kuntilla, about 50km north-west of Eilat, killing about a dozen soldiers, wounding six and taking 29 as prisoners in exchange for two own killed and two wounded. When EAF Vampires attempted to attack the withdrawing Israeli column, they were forced away by interceptors of the IDF/AF. In reaction to such Israeli attacks, Nasser ordered the deployment of Palestinian units against Israel: the Fedayeen were expanded into the 141st Battalion and ordered not only to conduct reconnaissance, but also to infiltrate, set up ambushes and plant mines. Even then, it took another Israeli outrage for the unit to actually be ordered into action: on 5 April 1956, the IDF mortared the centre of Gaza, massacring 56 civilians and wounding 103. The Fedayeen went into action and around a dozen Israelis were killed during the next few days, while on 12 April 1956 a pair of EAF Vampire FB.Mk 52s from the el-Arish based No. 31 Squadron penetrated Israeli airspace. Within minutes, both jets were intercepted by a pair of recently delivered, French-made Dassault Ouragan fighter-bombers, which hit the aircraft flown by Flight Lieutenant Lufti (serial number 1584), forcing him to make a belly landing near Abadat, a few kilometres south of Sde Boker. While Lufti spent nearly a year in captivity, his wingman withdrew back into Egyptian airspace.

NATIONALISATION OF THE SUEZ CANAL

Nasser's negotiations with the administration of US President Dwight D Eisenhower in Washington, and the government of Prime Minister Anthony Eden in London for financing of the High Nile Dam, had thus reached their high point amid a steady flow of reports about incidents along the ceasefire line and ever more Egyptian orders for Soviet arms. In February 1956, it was announced that the World Bank would lend US$500 million necessary for construction of the dam on condition of Britain and America lending a further US$70 million. Concerned that this would subject the Egyptian economy to Western control, Nasser requested several amendments to the draft agreement. Convinced that the Egyptian economy was shaky and Moscow could not replace them, Washington and London turned him down: the Eisenhower administration was under fierce Zionist pressure to drop the scheme entirely and, seeking a second term during elections to be held in November 1956, it was inadvisable for the US president to support a vast loan to aid a country considered to be veering towards the USSR. Thus, when on 19 July 1956 the Egyptian ambassador in Washington suggested that if the West would not help, Egypt would accept Soviet backing for the High Dam, the USA – followed by Great Britain, a day later – completely withdrew all their support.[8]

Left without a choice, on 26 July 1956 Nasser delivered a public speech from the same balcony in Alexandria from which he had evaded assassination, two years earlier: what an exultant crowd in front of him got to hear was something many Westerners could hardly believe. In a speech that revealed new and rather demagogic oratory skills, Nasser cemented his leadership of the Egyptian people by announcing that the Suez Canal would be nationalised, and that its GB£35 million annual revenue would be devoted to building the High Dam. Even before the Egyptian President had finished speaking, the Egyptian Army quietly moved into the Suez Canal Zone and took over the offices and installations of the Suez Canal Company.

By coincidence, the news of Nasser's speech arrived in London just as Eden was holding a dinner party for King Faysal of Iraq and his Prime Minister, Nuri as-Said. Said's reaction was to encourage the British Prime Minister to respond with resolution. However, he also warned about the danger of Britain allying herself with France and Israel to destroy Nasser, because that would have disastrous effects for Anglo-Arab relations. Tragically, Said's prophetical warning was ignored. On the contrary, henceforth securing the 'impudent' Egyptian President's overthrow became an obsession with Eden.

During the following days, fierce accusations about the nationalisation being illegal, as threatening the functional viability of the Suez Canal and thus the safety of European oil supplies and similar issues were published by much of the Western media. Actually, the take-over was legal, because compensation had been not only promised, but provided. It was effective, because the new national Suez Canal Authority soon proved that it could manage and operate the Canal even better than the old colonial company. Indeed, increased traffic volume in the weeks after the take-over assured that the Europe's oil lifeline remained undisturbed. The actual problem was that the British government felt humiliated.[9] Therefore, on 31 July 1956, Eden announced that Britain was, 'strengthening her forces in the eastern Mediterranean, as a precautionary measure, and that France had been given permission to station troops in Cyprus'. British commanders in the Mediterranean had also been alerted for possible military action, and the Chief of Staff was ordered to produce a study of the forces required to seize back the Canal. Although many of the best French units were already tied down in Algeria, Paris keenly signalled its preparedness to participate in a joint military intervention. Nasser's gesture had similar, unhappy effects in the USA, where representatives of the Eisenhower administration signalled their sympathies with Britain and France, and a preparedness to go along with their allies. It was under these circumstances that London and Paris began preparing their military intervention in Egypt, which subsequently also involved Israel, and resulted in a tripartite aggression.

BUILDING-UP THE FLEET

With hindsight, it can only be concluded that most reactions to the Czechoslovak Arms Deal – whether by Great Britain, Israel or France, were not only wildly exaggerated, but also extremely short sighted. Certainly enough, the sudden availability of large numbers of Soviet-made jet fighters, jet bombers and associated equipment, tools, spares and ammunition allowed a massive expansion of the EAF. However, this process proved far more problematic than expected in Cairo: indeed, it proved more problematic than conversion of the first two EAF units to MiGs. The reason was not only related to years of the training syllabus being interrupted by British embargoes and refusals to supply at last advanced training arms, but purges in 1952-1953: while the Air Force had nearly twice the number of necessary pilots, and at least a sufficient number of technicians, it lacked in terms of organisation, and the mass of its personnel simply had no experience on jet-propelled aircraft.

This became clear when, in late 1955 – as the first two MiG units entered the final phase of their build-up – the EAF made the decision to split No. 30 Squadron to provide a core for the reorganised No. 20 Squadron, which was to become the third to operate MiGs. To work up the latter unit, the Egyptians were forced to add an additional group of pilots and ground personnel, all of whom had only a little or no experience on jets. Even bigger problems were experienced during the conversion of Egyptian bomber crews to jet-powered Il-28s: as of 1955 the remaining Halifaxes and Lancasters of Nos. 8 and 9 Squadrons had been non-operational for years. In an attempt to make the conversion easier, the Egyptians and their Soviet advisors decided to temporarily merge the two units. Although the resulting bomber squadron slowly came into through early 1956, the progress remained slow and only a handful of crews were considered operational on the Il-28 even months later. Unsurprisingly, this – the second phase – of the EAF's conversion to Soviet-designed jet fighters and bombers proved less successful than the first one. Just how much so became obvious from reports about numerous accidents that occurred during the first months of 1956: while two, perhaps three MiG-15s and one Il-28 had been written off or badly damaged in earlier accidents, reportedly up to 14 or 15 'MiGs and Ilyushins' were damaged or written off through 1956.[10]

Egyptian and Czechoslovak ground crews conducting maintenance on the Il-28 coded 'K'. (Nour Bardai Collection)

A crew of three (including pilot, navigator/bombardier and tail-gunner) approaching an Il-28U conversion trainer – appropriately coded 'I' in black below the pilot's cockpit – shortly before their next training flight. (Nour Bardai Collection)

A flight of five Il-28s of the EAF seen low over Cairo in September 1956. (Nour Bardai Collection)

The highest attrition rate was not experienced by pilots converting to MiGs or Ilyushins, but by novice pilots training to fly Yakovlev Yak-11 training aircraft, which proved mechanically unreliable and sluggish in operation. Their deliveries began only in August 1956, but three Yaks are known to have been written off in accidents by late October, including two crashed by instructors during a rehearsal for a grand flypast planned for President Nasser in September 1956.[11]

Sadly, clear photographs of the first six MiG-15UTIs of the UARAF remain unavailable. This poor, but rare shot shows one example delivered immediately after the Suez War of 1956. This two-seat conversion trainer was in much demand during the wholesale conversion of the EAF to Soviet-made jets in 1956. (Nour Bardai Collection)

MEAGRE RESULTS

It was for similar reasons that the conversion of Syrian personnel to MiGs initially progressed very slowly. Instead of Syria, the first 25 aircraft were delivered to Egypt because the work on runway extensions at al-Mezzeh and Hama ABs in Syria was still underway, while Czechoslovak experts found Nayrab Airport unsuitable and recommended a complete reconstruction and expansion of the nearby Kweres and Rasin el-Aboud (not used since the times of the French mandate). Moreover, the Syrian air bases were found lacking serious air defences, and the SyAAF had no early warning radars whatsoever: all of the necessary equipment first had to be ordered and delivered, whether from Czechoslovakia or the USSR. Finally, it appeared simpler for Prague to deploy additional instructors to Egypt to supervise the conversion of a group of SyAAF pilots and ground personnel already undergoing training at the Air Force College in Bilbeis: this was why the Syrian MiGs were unloaded at Alexandria and then trucked to Abu Suweir AB in autumn 1955. In theory, this brought the total number of MiG-15s delivered to Egypt by early 1956 to no fewer than 116 airframes, further increased to 131 when the second batch of Syrian MiG-15bis arrived in Alexandria in October 1956.

In reality, because SyAAF personnel were still undergoing their training, the available Czechoslovak and Soviet instructors were already overtasked with converting Egyptians, and due to the lack of fuel and tools necessary to assemble newly delivered aircraft and then at least test-fly them – all caused by chaos in the EAF organisation – only 21 Syrian MiG-15bis and four MiG-15UTIs were ever assembled and test-flown. The remaining 20 aircraft were still in their transport containers, stored inside the hangars at Abu Suweir. Furthermore, while an additional team – the fourth – of Czechoslovak advisors arrived in Egypt in May 1956, and then provided ground courses to Syrian students, their flight training proceeded at a slow pace. The reasons were several and included language-related difficulties, and what the Czechoslovaks described as 'passivity' of Syrian students: their pilots refused to fly aircraft maintained by their own technicians because they still lacked experience, forcing the Czechoslovaks to supplement their Syrian colleagues on most of the work. Moreover, the Egyptians repeatedly failed to provide the necessary fuel, and one Czechoslovak instructor then crashed while test-flying a MiG-15UTI (actually a Czechoslovak-manufactured Avia CS.102, construction number 612792) near Dikhelia AB, and was killed. As a consequence, although the Syrians managed to complete their basic training on MiGs, they were still not combat ready as of October 1956. Without Syrian jets, and following training-related attrition mentioned above, only 64 MiG-15bis and MiG-15UTIs were available to the three EAF squadrons and one operational training unit as of late October 1956.

3
SUEZ WAR

The last British troops withdrew from the Suez Canal Zone in June 1956. In celebration, the Operational Training Unit of the EAF put up a large flypast over the Egyptian troops who took over the former British bases, including 48 aircraft, nine of which were MiG-15s. At the time, the pace of Egyptian training operations remained routine: it was only through July and August that the Main Directorate of Military Intelligence of Egypt (DMI) began receiving reports about the Anglo-French military build-up on Cyprus. Initially at least, Nasser and other members of the RCC found this hard to believe. On the contrary, after receiving several warnings about the planned Israeli invasion of Sinai, they became preoccupied with developments along the armistice lines and reinforced the Egyptian Army troop strength in the peninsula to about 45,000. Even then, all the Egyptian planning was purely defensive by nature: under the plan developed by the General Command, the Army was to deploy only a light protective screen along the armistice lines, while holding the majority of its units back in well-fortified positions in the northern centre of Sinai: the objective was to let the Israelis attack and then lure them into killing grounds of Egyptian choice. Correspondingly, only the old defence system in a triangle between el-Arish, Rafah, and Abu Aweigla (also 'Abu Ageila') was reconstructed.[1]

Additional intelligence reports indicating a serious threat of an Anglo-French invasion from September 1956, prompted a dramatic change of this plan: with such a threat being of an existential nature for the survival of the entire RCC, the defence of the Suez Canal, Alexandria and Cairo received priority. Correspondingly, major Egyptian Army units were withdrawn form Sinai, where the troop strength was reduced to 30,000, and redeployed along the Canal and in the eastern Nile Delta.

FIGHTER CONTROL SYSTEM

A trio of MiG-17Fs seen in flight near the Pyramids of Giza in late October 1956. (David Nicolle Collection)

The threat of an Anglo-French invasion caught the EAF not even midway through its conversion to Soviet-made equipment. Knowing that the mass of its units was not combat ready, and under the impression of what its officers thought they knew about the British and French air power, the General Command defined the EAF a 'clear underdog' in the case of a major war and ordered the Air Force to disperse and safeguard its aircraft and pilots. Correspondingly, all the squadrons received orders to prepare dispersal sites in the Nile Delta. Ironically, at the same time the EAF failed to pay attention to one of the crucial issues: the poor condition of most of its air bases. During the mid-1950s, the Egyptians had taken over many ex-RAF bases: most of these were not only found to be in a decrepit state but required significant improvements to enable jet operations. While by October 1956 the EAF had at least one runway stretched to 3,000m (9,800ft), there were next to no dispersal areas, emergency runways, or hardened aircraft shelters and maintenance facilities were almost non-existent. Indeed, some air bases were used in an unexpected fashion: Abu Suweir, for example, was the largest air base in Egypt at the time, yet primarily served the purpose of storing surplus aircraft. In turn, el-Arish remained the only air base with a concrete runway suitable for jet operations in Sinai: while some work was undertaken on other airfields on the peninsula – like el-Sur (Jebel Libni), Abu Aweigla, Bir Thamada, Bir Hama and Bir Gifgafa, none received a hardened runway by October 1956.

Perhaps most important was that even as of September 1956, the EAF still had no operational early warning radar network. Certainly enough, Egypt had a total of 12 radar systems of British and French origin, but most were non-operational for lack of spares and associated equipment. This is why their crews regularly turned down the few operational examples around noon every day. Supported by the Czechoslovaks, the Egyptians scrambled to repair some of the equipment, while the Soviets rushed 16 P-8 early warning radars (ASCC/NATO-codename 'Knife Rest A') – with a detection range of 150km (81nm) – to Egypt, together with more than 400 air defence guns and 14 fire-control radars (the latter to bolster protection of the EAF bases).[2] Although most of the equipment could not be brought into position in time, nor its personnel trained, some of the new

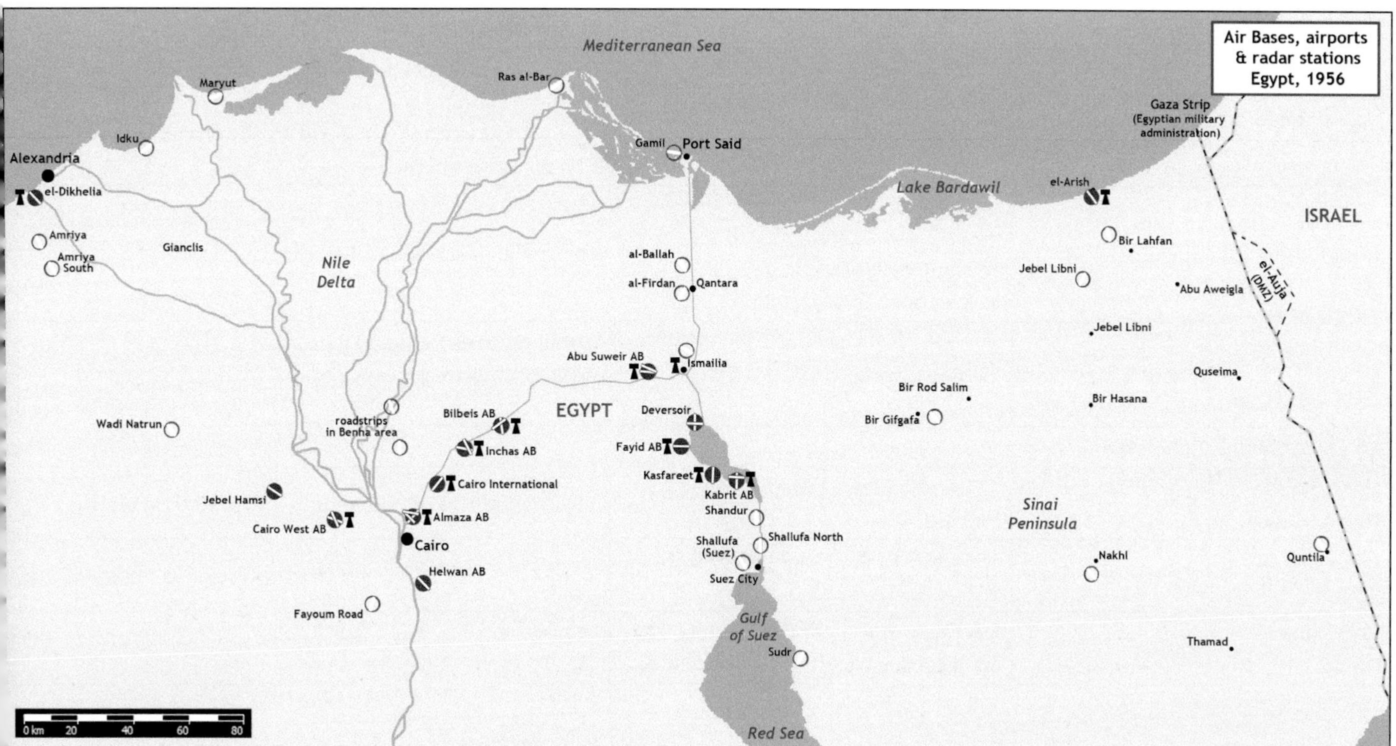

A map based on a scan from a Czechoslovak document providing a 'blueprint' of the Egyptian network of early warning radar stations as deployed before and during the Suez War. Black symbols denote radar stations deployed next to major air bases. (Czech National Archive, via Martin Smisek)

Table 2: EAF Order of Battle, 29 October 1956[7]			
Unit	**Base**	**Equipment**	**Remarks**
Chief of Staff EAF: Air Marshal Mahmoud Sidqi Mahmoud **Deputy Chief of Staff: Air Vice-Marshal Gamal Afifi** **Deputy Fighter Operations: Air Vice-Marshal Ali Attia**			
EAF Central Region, HQ at Almaza AB			
No. 1 Squadron	Almaza	18 MiG-15bis & 12 MiG-17F	CO Sqn Ldr Hinnawy, operational on MiGs, some aircraft at Kabrit
No. 2 Squadron	Cairo West	20 Vampire FB.Mk 52	CO Wg Cdr Kamal Zaki; 12-14 operational, some aircraft at Fayid; 15 additional Vampires stored at Abu Suweir
No. 3 Squadron	Almaza	11 C-47/Dakota	up to eight aircraft in reserve
No. 4 Squadron	Dikhelia	seven C-45	acting as Navigational School; all seven aircraft in reserve
No. 7 Squadron	Almaza	20 C-46	additional aircraft in reserve; combined with former No. 12 Squadron
No. 8 Squadron	Inchas, Luxor	12 Il-28	CO Wg Cdr Kamal Zaki; only four crews qualified; two aircraft at Cairo West, effectively combined with No. 9 Squadron; 10 aircraft stored at Luxor
No. 9 Squadron	Inchas	12 Il-28	CO Wg Cdr Hamid Abdel-Ghafar; only four crews qualified; effectively combined with No. 8 Squadron
No. 10 Squadron	Almaza	five Meteor NF.Mk 13	CO Sqn Ldr Salah ad-Din Husayn; only two operational
No. 11 Squadron	Almaza	four C-47/Dakota & Il-14; two Mallard, One S.51	
Il-28 OUT	Luxor	20 Il-28s & Il-28Us	
EAF Eastern Region, HQ Ismailia AB			
No. 5 Squadron	Fayid	eight Meteor F.Mk 8, one Meteor T.Mk 7	CO Sqn Ldr Mohammed Hilmi; seven F.Mk 4 and 17 F.Mk 4 in storage
No. 20 Squadron	Cairo West and Kabrit	12 MiG-15bis	CO Sqn Ldr Mohammed Nabil al-Masry; in the process of conversion to MiGs
No. 30 Squadron	Abu Suweir	15 MiG-15bis	CO Sqn Ldr Nazih Khalifa; operational on MiGs, some aircraft at Abu Suweir
No. 31 Squadron	el-Arish	-	CO Sqn Ldr Bahgat Hassan Hilmi; unit withdrawn from el-Arish; Vampires sold to Saudi Arabia, pilots undergoing ground course on MiG-15s and mostly served with No. 40 FTU during Suez War
No. 40 FTU	Fayid	eight Meteor F.Mk 4, three Meteor T.Mk 7; 10 Vampire FB.Mk 52, four Vampire T.Mk 55; four Harvard	CO Sqn Ldr Salah ad-Din Husayn
MiG OUT	Kabrit	12 MiG-15bis & 10 MiG-15UTIs	further 20 MiG-15bis and four MiG-15UTI ordered by Syria were assembled but crews lacked operational training course
Air Force College, Bilbeis AB			
Elementary Flying School	Bilbeis	13 Gomhouriya, 51 Chipmunk, 36 T-6/ Harvard	13 Gomhouriyas, 37 Chipmunks, 19 Harvards in reserve
Advanced Flying School	Bilbeis	15 Spitfire F.Mk 22, eight Fury FB.Mk 11, 20 Fiat G.55, seven Yak-11	
Agricultural Flight	Bilbeis	five BT.13, one Super Cub, 22 Sokol, 22 Magister, two Storch	

radars were deployed at carefully selected sites and linked-up into a network with the existing ones in late October 1956. Thus came into being the Fighter Control System: commanded by Air Vice-Marshal Ali Attia, this included three radar stations deployed in the Nile Delta (at Nomman, el-Agami and Dikhelia), one outside Cairo (at Zamalek), two in the Suez Canal Zone (at Gamil and Abu'l-Sultan west of Deversoir), and three on the Sinai Peninsula. Four of these were equipped with French-made ESV2s and five with Soviet P-8 systems.[3]

Because most of this work was undertaken in the last week before the Suez War of 1956, its results entirely escaped the attention of Israeli and Western intelligence agencies. In turn, this resulted in Western assessments that the EAF's operational efficiency would be very poor.[4] Certainly enough, the situation was far from satisfactory even in the opinion of Czechoslovak advisors: indeed, they concluded it as 'critical'. They declared only Nos. 1 and 30 Squadrons as operational in July 1956, and even then the conclusion was that Egyptian pilots were still in need of tactical training. Correspondingly, instructors took care to provide advice in air-to-air and air-to-ground gunnery, and operations in pairs, sections of four and then larger formations of aircraft. Arguably, because this part of the training was undertaken in great hurry, and due to the sheer number of EAF and SyAAF pilots in need of training, the Czechoslovaks were not only completely exhausted by all the work but considered it insufficient to make their Egyptian students combat ready. Indeed, the first operational exercise, undertaken on 26 and 27 September 1956, was considered an abysmal failure – not only due to the peculiarities of the EAF's pilots (many of whom were found lacking the necessary discipline and preparedness to follow advice of their instructors), but also because the Fighter Control System was still non-operational by that time.[5]

Eventually, according to Egyptian records, the EAF reached a personnel strength of about 6,400 – including about 440 pilots – by October 1956: while only about 120 of the latter were qualified to fly jets, and about 40 others considered ready for combat on older types, between 250 and 260 were about to undergo conversion courses in Czechoslovakia and the USSR. Furthermore, while the Air Force entered the Suez War with only 76 operational aircraft, and a similar number in reserve, the 12 MiG-17Fs from the April 1956 order arrived in Alexandria in late October, and another batch was expected for November. Quickly assembled by Soviet advisors, some of these brand-new aircraft were integrated into No. 1 Squadron and six pilots converted to them.[6]

SOVIET ADVISORS

The presence of what were generally seen as 'Soviet' advisors in Egypt of 1956 was an issue of major concern for Western powers and Israel during their planning for the tripartite aggression on Egypt. Due to experiences from the recent Korean War – and because the French deployed 36 combat aircraft to bolster Israeli air defences – the assumption emerged that the USSR might deploy its troops, 'probably disguised as volunteers' for combat purposes in Egypt. In reality, caught off-guard by the Czech Arms Deal, the intelligence branch of the IDF, AMAN, was keen not to experience another mishap and thus tended to provide 'worst case' scenarios to its supervisors. Moreover, AMAN supposedly intercepted 'evidence' that Nasser did raise the issue in Moscow as early as of March 1956. Accordingly, on behalf of the Egypt, Saudi Arabia and Syria, he apparently requested, 'Muslims from the Soviet republics in Central Asia [to] assist them when necessary in using military technology…"[8]

Exactly why a president of Egypt renowned for persecution of Islamists demand 'Muslims' from the USSR to fight for him as of March 1956 remains unclear and appears rather doubtful. Moreover, even the same AMAN report stressed that Moscow replied negatively. Nevertheless, bolstered by widespread arrogance and prejudice about Arabs, such Israeli assumptions spread west, regardless of how unsubstantiated. Indeed, the idea gave birth to a mass of legends about direct Soviet involvement in the 'Suez Crisis' as the following conflict is recalled in the West, and later on.

In reality, because the entire Czech Arms Deal was officially presented as a contract between Egypt and Czechoslovakia alone, the presence of Soviet advisors in Egypt during 1955-1956 remained very limited. The General Staff of the Soviet Armed Forces (GenStab) deployed only a small group of officers all carefully screened by – and including officer of – the Main Intelligence Directorate of the Soviet Armed Forces (*Glavnoye razvedyvate'lnoye upravleniye*, GRU). Their presence was considered so sensitive that the Soviets were always officially declared as Czechoslovakian citizens. This reached a degree where older EAF officers remain convinced that in 1955 and 1956 they were being trained by Czechoslovak personnel only. Immediately upon the onset of the Franco-British invasion, all Czechoslovak and Soviet advisors were withdrawn and none of them flew any kind of combat operations.[9]

FIRST SIGNS OF TROUBLE

For the Egyptian pilots freshly converted to their brand-new MiGs and Ilyushins, the first signs of trouble emerged on 29 October 1956, when two piston-powered de Havilland Mosquito reconnaissance fighters of the IDF/AF crossed the Suez Canal Zone. A few hours later, a pair of French-made Dassault Mystère IVA fighter jets was detected while passing at low altitude over the Mitla Defile, in western central Sinai. Powered up, early that afternoon EAF radars reported intensive Israeli activity along the armistice line, followed by a third reconnaissance mission over the Canal Zone. With such Israeli violations of the Egyptian airspace now being a daily occurrence, the Egyptian authorities did not assume anything sinister was brewing, and thus there was no immediate reaction: EAF interceptors were scrambled only once the aircraft carrying the US Ambassador to Egypt strayed over a military zone east of Cairo. Led by Squadron Leader Hilmi, Alaa Barakat thus flew his first operational sortie ever – in a Meteor F.Mk 8 of No. 5 Squadron – from Fayid AB. The two Egyptians obliged the US aircraft to land at their base without further problems.

In fact, it was only later in the afternoon, when a dozen Israeli Mystères appeared high above Kabrit AB with the intention of intercepting any EAF fighters that could have taken off to intercept a formation of 16 low-flying Douglas C-47 transports carrying paratroopers to the Mitla Defile, that the Egyptians noticed anything unusual. Although visible from the ground, the Mystères were not immediately recognised as enemy aircraft because, from a distance, they looked similar to MiG-15s. Unknown to the Egyptians at that point in time was that they were able to fly 'undetected' all the way to Kabrit because during the afternoon several North American P-51 Mustang fighters of the IDF/AF had cut vital telephone wires in Sinai, thus interrupting communications between outposts on the ceasefire line with the headquarters in Cairo.

Further confusion was caused later that evening, first by the news that contact had been lost with an Il-14 transport underway from Damascus via Cyprus to Cairo. Piloted by Squadron Leader Mustafa Mahmud Hilmi Ismail, with Squadron Leader Kamal ad-Din Ahmad Abu l'-Rahim as co-pilot, the aircraft wore the serial number 1101,

This was the Il-14 with serial number 1101. Specially equipped as a VIP-transport, it was left in bare metal overall. The aircraft was shot down by an Israeli Meteor NF.Mk 13 late in the evening of 29 October 1956. (David Nicolle Collection)

An elevated view of another Il-14 of the EAF as of 1956-1958. Generally ignored at the time, 20 aircraft of this type were soon to form the backbone of the Egyptian transport fleet. Notable on this example is the application of the service title (in Arabic and English), and the serial number 1105, on the rear fuselage and repeated, 'RAF-style' on undersurfaces of the wing. Another interesting detail is that the 'last two' of the serial were frequently applied in white on a black circle on the nose. (Nour Bardai Collection)

and was the example equipped as a VIP-transport, donated to Nasser by the Soviets. On 25 October, it took a delegation led by Field Marshal Amer to Jordan. Amer's next destination was Damascus but it was considered too dangerous to fly there from Amman, because the Arabs feared an Israeli interception attempt. Therefore, Amer and his team travelled by road. On the evening of 29 October, the Il-14 flew back to Cairo, but instead of carrying Amer, there were 16 crewmembers and passengers on board, including several foreign journalists. As became known only decades later, the aircraft was indeed intercepted and shot down by a single Israeli Meteor NF.Mk 13 night-fighter in a blatant act of state-sponsored terrorism: the Israelis intended to assassinate Field Marshal Amer, but – entirely unaware of the tragic fate of the Il-14 – the lucky Egyptian Minister of War left Damascus only later that night, and then using a Dakota piloted by Flight Lieutenant Sa'ad ad-Din Sharif.[10]

FIRST CONFUSION, THEN SUCCESS

Meanwhile, the Egyptian Army reported the landing of Israeli paratroopers east of the Heitan Defile, the eastern entrance to the Mitla Defile. This report, and another one about an IDF column that crossed the armistice line and advanced in the direction of Thamad, confused the General Command in Cairo and prompted an urgent meeting with the RCC at the Joint Headquarters in Heliopolis. The situation was confusing because the Egyptians were not certain if the Israelis were launching an invasion, or another raid. Unaware of the plan for a tripartite aggression, they drew the wrong conclusion:

that Israel was acting on its own because Britain and France had decided not to attack.[11] Correspondingly, the General Command activated the plan for defence from an Israeli invasion and began dispatching reinforcements into Sinai. Field Marshal Amer ordered the commander of the EAF, Air Vice-Marshal Sidqi Mahmoud Sidqi, to re-deploy all available aircraft to air bases closer to Sinai and to launch air strikes on the Israeli paratroopers. Because the EAF's Il-28s were still not combat ready, the Air Force failed to fly any air strikes during the night from 29 to 30 October. However, the Fighter Control System of the EAF was up and running and thus, when the government in London ordered 'last minute' high altitude reconnaissance of Egyptian air bases prior to its planned intervention, the British experienced a major surprise. All four of the Canberra PR.Mk 7s involved were intercepted, although underway in dark pre-dawn skies at altitudes between 10,000 and 13,000m (32,808-42,650ft). Indeed, with the Fighter Control System's radars picking up incoming targets in time, four MiG-15bis of No. 1 Squadron were scrambled and vectored to intercept. Most successful was the one flown by Flight Lieutenant Sayd al-Qadi, who claimed a Canberra as damaged or shot down. Actually, his target – coded WT540, was seriously damaged by a hit to the port elevator. However, on return to base, Qadi's post-mission report was declared 'untrustworthy' and the pilot reprimanded 'for lying' by his superiors. This is the most likely reason why it remains unknown who the pilot of the second successful MiG-15bis was that morning, whose shells narrowly missed another British reconnaissance bomber (Canberra PR.Mk 7 coded WH801). Although not resulting in the downing of any of the RAF aircraft, this Egyptian operation was highly successful in forcing the British to postpone the start of their invasion to the following night, and then to order their bomber crews to fly at unusually high altitudes, in turn greatly decreasing their precision.

Although having very little time to work up its MiG-15 fleet before the coming war, the EAF did take care for some of these to be equipped with underwing launch rails for 80-88mm unguided rockets of Hispano-Suisa design. This bottom view of a MiG-15bis nicely shows the position of the rails in question (see circle hand drawn onto the original photograph). These enabled Hinnawy and his wingmen to carry bombs and rockets during their first sortie on the morning of 30 October 1956. (Nour Bardai Collection)

FRACAS OVER MITLA

The Canberras were hardly outside the Egyptian airspace, when the EAF went into action against the Israelis east of the Heitan Defile. At sunrise, four Vampires flew an armed reconnaissance mission over the area but failed to detect the enemy due to the thick fog. Instead, they flew on for nearly 150 kilometres south-east to find, on the way back, the Israeli column approaching the Quntila area. At 09.15hrs, as the fog cleared, No. 1 Squadron launched two formations of MiG-15s. The first included four jets and was led by that unit's CO, Squadron Leader Hinnaway, and attacked the Israelis near the Heitan Defile. This strike took the paratroopers by surprise, and not only wounded four, but also destroyed a Piper L-8 Cub of the IDF/AF on the ground. The second formation consisted of six MiGs and surprised the Israeli column near Thamad. Sweeping low, the Egyptians set six vehicles ablaze and wounded a number of troops. The Israelis were barely able to evacuate their casualties, when their column was hit again, this time by four MiG-15s, around 10.00hrs. This exercise was repeated about an hour later, when four Vampires – escorted by two MiGs – attacked the paras at Heitan, while four Meteors rocketed the column near Thamad, this time destroying a truck and a mortar, plus three other vehicles: indeed, by this time the Israeli formation was scattered to a degree where it took two hours to reassemble and start moving again.

By this time, interceptors of the IDF/AF were flying combat air patrols (CAPs) over Sinai but remained at high altitude. As the casualties on the ground began to mount, they were ordered down low and to engage Egyptian fighter-bombers and ground troops. Their first strikes hit a column of the Egyptian Army approaching

Wreckage of the Israeli Piper L-8 destroyed by Hinnawy and his wingmen near the Heitan Defile on the morning of 30 October 1956. This was the first enemy aircraft destroyed by an Arab MiG. (IDF, via Albert Grandolini)

the Mitla Defile from the west, causing it losses in vehicles, even if none in personnel. The EAF MiGs were airborne but failed to intervene. An operational pattern had now developed in which the EAF Vampire-units operated against the relatively close Mitla area, while Meteors engaged targets farther away: both were escorted by MiGs, which also flew ground strikes. One exception was an attack by six MiGs and four Meteors on the Israeli paratroopers now advancing through the Heitan Defile, which once again left numerous vehicles ablaze. It was only at this point in time that the first clash between Egyptian and Israeli fighter jets took place: unsurprisingly considering all of the pilots were unused to the high speed of enemy aircraft, this ended inconclusively. Shortly after, Hinnawy flew the first operational sortie in a MiG-17F of No.1 Squadron, when – around 15.00hrs – he was scrambled solo to intercept a four-engined British aircraft underway north of Port Said. Although reaching a favourable position behind his target inside Egyptian airspace, he was denied permission to open fire. The ground controller advised Hinnawy, 'Don't fire because the English have not declared a war!' [12]

ISRAELI 'ATTACK' ON KABRIT

The unexpectedly strong EAF activity made the Israelis extremely nervous. Around 15.00hrs, the AMAN received an – erroneous – intelligence report indicating that the Egyptians were about to launch a massive attack on the Israeli paras in Heitan with no fewer than 24 fighter jets. Scrambling everything they could, the IDF/AF directed all available Mystères over the Mitla Defile – and beyond. Of course, these encountered no MiGs and thus several formations continued all the way to Kabrit AB. Nine Israeli jets reached the Egyptian air base just as two sections, each of four MiG-15s, were about to launch another air strike: they immediately dived to attack the first pair of EAF fighters. The Israelis missed, and both MiGs accelerated away. However, another pair of Israelis then hit the MiG piloted by Sobhi at-Tawil shortly after he lifted off. The Egyptian pilot ejected. In turn, although all the time busy evading the more numerous Israelis, the leader of the first Egyptian pair, Flight Lieutenant Hussayn Sidqi, claimed to have hit the wing of the Israeli that had shot down Tawil. Ultimately, the air combat ended with the Israelis claiming a MiG-15 and a MiG-17 (one 'flown by a Soviet advisor') as shot down, without loss: the damaged Mystère returned safely to Hatzor AB. The Egyptians claimed one possible for one loss, while a related Czechoslovak report concluded that two Mystères were shot down in exchange for one MiG-15bis.[13]

Once the Israelis were away, Nazih Khalifa was free to lead a quartet of MiG-15s from No. 30 Squadron into a new attack on the Israeli column in the Thamad area. After setting several ground vehicles afire, his formation then ran into a lonesome Piper Cub of the IDF/AF. The veteran Israeli pilot waved and bobbed his aircraft for several minutes, but in vain. Eventually, he was shot down by gunfire from the leader of the Egyptian formation.

'STABILISING' SITUATION

Still uncertain about Israeli intentions, the General Command in Cairo remained uncertain how to react. However, what the Egyptians understood was an 'IDF/AF attack on Kabrit', was considered a serious provocation and thus No. 8/9 Squadron EAF received the order to counterattack. Contrary to what the Israelis did as soon as they had acquired Boeing B-17 bombers from the USA back in 1948, and also to countless Israeli reports that Egypt was buying Il-28s to 'bomb the Israeli population', the Egyptians never even considered acting that way during this nor any other of the Arab-Israeli Wars. Instead, the two crews that eventually flew this mission received orders to attack air bases known in Egypt as Qastina (Hatzor in Israel), and Aqir (Tel Nov). According to unofficial Egyptian sources both reached their targets, and their crews reported causing large fires: the Israelis stress just one stick of bombs was released and fell in the Ramat Razi'el area, east of Jerusalem.[14]

Overall, the Egyptians – and particularly the EAF – looked back at 30 October 1956, the first day of serious fighting of this war, with considerable satisfaction. The EAF had launched slightly over 50

Squadron Leader Nazih Khalifa, CO of No. 30 Squadron – only the second Egyptian unit to convert to MiG-15s – and the most successful fighter pilot of the EAF during the Suez War of 1956. (Gamal al-Khalifa Collection)

A MiG-17F scrambling in full afterburner. Gauging by the size of the roundel on the rear fuselage, this photograph shows one of the aircraft delivered after the Suez War, though the scene is reminiscent of many scrambles during that conflict. (David Nicolle Collection)

combat sorties, losing only one MiG-15 in the process. Because most of the pilots involved were still fresh from their tactical training, and their Czechoslovak and Soviet advisors emphasised organisation and safety over aggressiveness, and because of their slow-firing guns, they performed less successfully in air combat. Still, the overall score was 2:1, because the EAF claimed an Israeli Mystère and a Piper Cub, and destroyed another Cub on the ground, while losing one MiG. Moreover, Egyptian pilots proved highly effective in ground attacks: their air strikes pinned down the Israeli parachute battalion near Heitan Defile for hours, and repeatedly stopped and dispersed the advance of the column underway from Quntila via Thamad, destroying 23 of its vehicles. Elsewhere in Sinai, the first Israeli ground assault on Abu Aweigla was thrown back in confusion and only served the purpose of making the seriousness of the Israeli invasion obvious in Cairo. Finally, the Egyptian ground forces claimed numerous Israeli fighter-bombers as shot down during the day, before the battle even began in earnest, and reinforcements were still streaming over the Suez Canal. Quite unsurprisingly, the General Command thus concluded the situation to be 'stable' and everybody was looking forward to even more success on the next day.

ANGLO-FRENCH ULTIMATUM

It was at that moment that – at 19.00hrs Cairo Time – Britain and France issued the ultimatum for Egypt and Israel to evacuate an area 16km (10 miles) to either side of the Suez Canal, supposedly in order to safeguard the normal function of the crucial waterway. For all practical purposes, the Israeli reaction was irrelevant: the IDF had not a single soldier that close to the Canal. Unsurprisingly, the Egyptians rejected such an ultimatum: however, it now dawned not only on the RCC and the General Command, but also upon Nasser that they were facing an aggression by three powers, not just Israel. This led to a major dispute between Nasser and Amer over the issue of what to do next: the President insisted on a complete withdrawal of all units from Sinai, while Amer wanted to continue fighting Israel. Nasser won through a political rather than military argument, concluding that the entire purpose of any invasion was to occupy the Suez Canal Zone, which in turn would cut off the best Army units to the east of it: he insisted on the concentration of all available forces in the Port Said and Ismailia areas, and on the EAF being grounded to avoid suffering personnel losses in a frontal collision with the superior air forces of Britain and France. At least as important was that while officially accepting this decision, Amer did everything possible to delay the withdrawal from Sinai. Indeed, for most of 31 October, the two best Egyptian armoured brigades continued converging on Bir Gifgafa and Bir Rod Salim, following the pre-war defence plan. Moreover, the EAF continued fighting the Israelis: while less experienced crews were used to transfer 20 reserve MiG-15s to dispersal airfields set up in the Nile Delta, and to evacuate nine Il-28s and one Il-14 via Luxor to Saudi Arabia, all pilots considered combat ready were ordered to join operational units, where ground crews were to prepare all the available aircraft for operations.[15]

CRUCIAL DELAY

31 October 1956 began almost the same way as the previous day: with foreign reconnaissance aircraft violating Egyptian airspace. The primary difference was that this time not only high-flying Canberras of the RAF, but also low-flying Republic RF-84F Thunderflash reconnaissance fighters of the Armée de l'Air (French Air Force, AdA) were involved. The EAF scrambled to intercept every single one, but this time only two of them managed to catch up with a solitary French jet: the first indication for the French pilot that he was being pursued was when tracer began whizzing past his cockpit. He aborted his mission and returned to Cyprus. In this fashion, the AdA and the RAF obtained crucially important photographs informing them that the EAF had over 110 operational MiG-15s, 14 Meteors, 44 Vampires and 48 Il-28 bombers available, deployed as follows:

- Cairo West: nine Vampires, 16 Il-28s
- Almaza: 25 MiG-15/17, four Meteors, 21 Vampires, 10 Il-28s
- Inchas: 20 MiG-15s
- Abu Suweir: 35 MiG-15s
- Kabrit: 31 MiG-15s
- Fayd: nine Meteors, 12 Vampires
- Kasfareet: oneMeteor, two Vampires
- Luxor: 22 Il-28s.

Conscious of interceptions of high-flying Canberras 24-hours before, the British and the French concluded that the Egyptian Air Force was far more dangerous than expected, and thus decided to postpone their offensive on EAF air bases until the following evening. This delay not only enabled the Czechoslovaks and Soviets to evacuate their personnel from Egypt, but also enabled the EAF to continue evacuating its aircraft: all but eight Il-28s had been flown to Luxor, followed by 10 MiG-15bis and at least three, possibly four SyAAF MiG-15UTIs.

ALL-OUT EFFORT

Because once the Anglo-French ultimatum expired around 06.00hrs of 31 October, there was no immediate onslaught on its air bases, the EAF scrambled its operational aircraft into an all-out attack on the Israeli paratroopers on Sinai. The latest information provided by the Army was that the column moving from Quntila via Thamad had joined the paras in the Heitan Defile. Therefore, the commander of No. 40 Squadron at Fayid rushed four of his pilots into the air, intending to hit the Israelis on the ground before the first of the expected IDF/AF Mystère patrols could reach the area. Indeed, armed with unguided rockets and ammunition for their four internal 20mm guns, the four jets led by Squadron Leader Bahgat Hilmi pressed their attack home despite the morning mist, each making one pass. However, while preparing for their second strafing run, they came under attack from at least two Israeli interceptors. The rearmost two Vampires in the column were hit in quick succession and caught fire. Second in the formation, the jet piloted by Pilot Officer Gabr Ali Gabr was overshot by one of Mystères – enabling the Egyptian to fire a quick burst at the fleeting target – but the Israeli then went on to hit Hilmi's aircraft. While one Vampire spiralled out of control almost immediately, the two others crashed on the return west: two of their pilots were killed, while one was captured by the Israelis, leaving the young Gabr the sole survivor.[16]

The tragedy that befell No. 40 Squadron prompted a major change in modus operandi of the EAF: for the rest of that day, none of the older and slower jets were allowed to fly over Sinai without top cover by MiGs. Moreover, the Egyptians began operating in ever larger formations, usually proceeded by additional MiGs. About one hour later, a quartet of MiG-17s from Almaza AB flew a sweep over Mitla and promptly engaged a pair of Israeli Meteors but failed to bring any down. This clash remained an exception, then although the Israeli paras then launched an attack into the Defile, they were spared further Egyptian air strikes: instead, the EAF redirected its

MUSTAFA SHALABY EL-HINNAWY: TACTICIAN

Probably the most under-recognised of all the early Egyptian MiG pilots was Mustafa Shalaby el-Hinnawy, a crucial figure for the EAF during at least two periods. While nothing is known about his earlier life, or specific other periods of his military career, it is known that – according to the usual procedure at the time, in which future pilots first had to complete two years of military training before switching to flight training – Hinnawy joined the Military Academy in 1942, where Gamal Abdel Nasser was one of his instructors. He switched to the Air Force in 1947, graduated as the top of his class from the Air Force College a year later, and flew Spitfire F.Mk IXs during the Palestine War of 1948-1949. His next posting was that of a flight instructor at Almaza until 1953, when he was re-assigned to the FCU. Amongst others, he taught Hosni Mubarak to fly Spitfires and Furies to the verge of their envelope. In 1955, Hinnawy was appointed the commander (CO) of No. 1 Squadron – the first EAF unit equipped with the MiG-15bis, and then with MiG-17Fs. It was in this function that he flew and fought during the Suez War of 1956. Sadly, next to nothing is known about him over the following 10 years: most likely, Hinnawy underwent advanced staff and command courses at home and abroad, de facto exiled by Field Marshal Amer for his insistence on military professionalism. What is known is that in 1967 he wore the rank of Brigadier-General and was appointed the Head of Air Operations of the United Arab Command – a military command established in 1964, commanded by Egyptian Major-General Riyadh, envisaged to coordinate the work of all Arab armed forces in the case of a war with Israel, and headquartered in Amman in Jordan. At this post, he travelled to Jordan and Iraq shortly before the June 1967 War, but during that conflict, primarily due to a series of colossal failures and chaotic orders by Field Marshal Amer, never managed to organise and coordinate operations of the Iraqi Air Force (IrAF), SyAAF and the Royal Jordanian Air Force (RJAF) to any degree.

Shalaby el-Hinnawy around the time of the Palestine War in 1948-1949. (David Nicolle Collection)

On 5 July 1967, Hinnawy was appointed to the Enquiry Committee led by Major-General Hassan Mutawah, established to define the reasons for the catastrophe that befell the Egyptian armed forces during the June 1967 War with Israel. Following months of investigations, this Committee filed two reports: a short summary of 93 pages, including four concerning the Air Force, and a complete version, *Document 44*, containing over 1,000 pages (including over 50 about the Air Force). Both drew the same conclusion: that most of the shortcomings were known as early as of 1956-1963, but that due to the politicisation of the armed forces under Field Marshal Amer, and widespread corruption, patronage and nepotism, next to nothing was undertaken to rectify the problems or improve the overall condition of the armed forces.

Reputed to be an energetic officer, excellent tactician, organiser, and administrator, and certainly one of very few top Egyptian officers with full understanding of air power at the time, Hinnawy was appointed the commander of the Egyptian Air Force on 2 November 1967 and led it into a period of complete re-organisation. Amongst his most notable measures was the decision to fortify all the air bases: over 500 hardened aircraft shelters were constructed while he was in command and dominate Egyptian air bases until the present day. Another was the order for the Air Force to stop haphazardly engaging Israeli aircraft violating Egyptian airspace, and instead engage them in well-prepared operations undertaken by hand-picked pilots – tactics subsequently happily adopted by the Israelis. Yet another was the intensification of tactical training for Egyptian fighter pilots to up to 200 hours per annum.

Following several differences with superior officers, perhaps even with Nasser, Shalaby el-Hinnawy was – unceremoniously – dismissed from his position on 17 June 1969, supposedly as punishment for failing to prevent a pair of Israeli fighter-bombers from thundering down the full width of Cairo at low altitude. He spent the rest of his life in quiet retirement: unlike almost all other top Egyptian officers of this period, he left no published memoirs behind.

Hinnawy in discussion with other officers of his squadron. Visible to the left is Igor Bozhenko, a Soviet advisor assigned to No. 1 Squadron, EAF, during the Suez War of 1956. (David Nicolle Collection)

attention to supporting the 1st Armoured Brigade of the Army as it advanced from Bir Gifgafa farther east in an attempt to reinforce the defenders of the Abu Aweigla complex.

As far as can be reconstructed from several dozen different accounts, the first Egyptian formation to reach this area included eight MiG-15bis from Kabrit, which flew top cover for four Meteors from No. 5 Squadron and four Vampires from No. 31 Squadron. While fighter-bombers attacked their targets undisturbed, MiGs found no Israeli interceptors to fight, and it was only once they were on the way back to base that the section led by Squadron Leader Nazih Khalifa sighted a pair of Ouragans. Khalifa split his formation into two pairs and sandwiched the Israelis, provoking a dogfight. While missing in his first attack, the Egyptian formation leader turned around, and – protected by his Number 2, Flight Officer Kefi – approached to within 150 metres of the enemy leader, to hit the Ouragan's wing and cause it to spiral towards the ground. His numbers 3 and 4, Flight Lieutenant Badr and Flight Officer Zuhayr, meanwhile attacked the other Israeli, claiming that it exploded in the air.'[17]

According to the Israelis, only one IDF/AF jet was hit in this engagement before a pair of Mystères arrived to help extract them. In turn, one of the latter ran out of fuel while returning to Hatzor AB and made a belly landing: the aircraft was subsequently recovered and repaired. Nevertheless, and once again, the action for Khalifa's section was not yet over. On the way back west, the four MiGs bounced a pair of Israeli Mustangs, somewhere between Bir Gifgafa and Bir Hama, and Flight Lieutenant Zuhayr hit one, causing it to crash-land in the desert.[18]

While taken during a pre-war parade in Cairo, this photograph nicely illustrates what was going on in the skies over the Sinai Peninsula during the afternoon of 31 October 1956, when the EAF began operating large formations of MiG-15s. With too few pilots to counter them, the IDF/AF stayed away from challenging a numerically superior enemy, in turn confirming the Egyptian assessment that they were on the way to ascertaining aerial dominance over the battlefield. (Albert Grandolini Collection)

Around 13.00hrs, a section of MiG-15s intercepted another pair of Ouragans, and claimed another probable, but it seems that neither of the two Israeli jets was hit. Only minutes later, four MiG-15s from Abu Suweir clashed with a pair of Ouragans between Bir Hama and Bir Hasana, provoking a five-minute-long chaotic dogfight in which most of the Egyptians concentrated on attacking one of two Israelis. Eventually, Flying Officer Farouq el-Ghazzawy claimed one kill, and the EAF credited him with a probable (while the Israelis deny any losses). In turn, Ghazzawy's jet received slight damage to its cockpit and wing, but the pilot returned safely back to base'.[19]

Shortly afterwards, a section of MiG-15bis each from Abu Suweir and Kabrit was about to attack Israeli troops in the Abu Aweigla area, with jets from the former air base providing top cover for the latter, when they were caught by surprise by two pairs of Mystères, north of Bir Hasana. Pilot Officer Fouad Kamal saw the approaching Israelis, and attempted to warn others via radio, but nobody reacted: seconds later, his MiG was hit and fell into a spin,

Abd ar-Rahman Muharram's MiG-15bis (serial number 901 or, more likely, 1901), as found by the Israelis, ditched in Lake Bardawil. (IDF)

A nice study of Muharram's MiG-15bis against a backdrop of slightly larger (and faster) Mystère IVAs of the IDF/AF. The Israelis recovered the damaged jet but failed to make it flyable again. (IDF)

A gun camera sequence showing the final moments of the attack on Fouad Kamal's MiG-15bis on 31 October 1956. Several hits by 30mm shells knocked the jet into an irrecoverable spin, forcing Kamal to eject. Clearly visible are identification stripes around the wings and on the rear fuselage of the MiG. (IDF)

forcing Kamal to eject. He joined Egyptian troops to return safely back to the Suez Canal a few days later.[20]

Flying at high altitude, another Egyptian formation ranged all the way to el-Arish, which was now under sustained air strikes by the IDF/AF. However, the local radar had been knocked out and thus the section of four MiG-15bis was ambushed by four Mystères: the jet piloted by Pilot Officer Abd ar-Rahman Muharram received a hit in its starboard wing. Although managing to escape by flying at low altitude, Muharram eventually ran out of fuel and ditched in Lake Bardawil. Much later, the Israelis found his abandoned MiG, and took it away for testing and photographs of his jet have ever since appeared in most accounts of the Suez War of 1956. Around 15.30hrs, four MiG-15bis launched from Kabrit to provide top cover for four Meteors from No. 5 Squadron sent in the direction of Jebel Libni. Underway there, they detected the column of Israeli paratroopers that were withdrawing from the Heitan Defile and decided to attack it instead: numerous vehicles were set on fire.[21]

About an hour later, 14 MiG-15s from Abu Suweir and Kabrit attacked Israeli troops in the Abu Aweigla area and, after expending all their bombs and rockets, then scared away a pair of Mystère IVAs. This action convinced the Egyptians that they had achieved aerial dominance over the Sinai Peninsula. This conclusion was based not only upon the fact that its jets flew 120 combat sorties during the day and its pilots claimed six aerial victories (of which two can be considered as confirmed), but that they operated largely unmolested deep over Sinai. Moreover, the Egyptian Army reported causing heavy losses to the low-flying fighter-bombers of the IDF/AF, and also the first appearance of Republic F-84F Thunderjets of the AdA deployed in Israel. From the Egyptian point of view, there was no reason for such a French intervention if the Israeli Air Force was operating effectively. With hindsight, this might appear overoptimistic and – certainly enough – is fiercely disputed by the Israelis. However, contrary to what was subsequently reported about its performance during this war, the EAF was far from being seriously concerned about its position vis-à-vis the IDF/AF, and was anything but defeated, as often reported since.

AERIAL ONSLAUGHT

British bombing raids on Egypt began soon after nightfall on the evening of 31 October. Putting priority on knocking out four major Egyptian air bases – Abu Suweir, Almaza, Inchas and Kabrit – Vickers Valiant and Canberra bombers of the RAF saturated these with large numbers of 225kg and 400kg bombs (500lb and 1,000lb, respectively) released from altitudes up to 12,192m (40,000ft).[22] Cairo International Airport was also accidentally hit, despite the intention to spare it because it did not have any fighters, and because US citizens were evacuated along the nearby highway. Once again, the EAF put up a spirited – even if feeble – performance. At least two Meteor NF.Mk 13s are known to have been airborne that night: Squadron Leader Salah ed-Din Husayin, the CO of No. 10 Squadron (Egyptian sole night-fighter outfit) intercepted a Valiant from No. 148 Squadron, RAF, forcing it into violent evasive manoeuvring. Husayin later claimed to have actually scored a few hits, but the Egyptian propaganda machinery promptly claimed a kill, while the British bomber actually came away without any damage. The other Meteor NF. Mk 13 of the EAF intercepted a Canberra heading for Inchas, but the vastly superior performance of the bomber at high altitude enabled its crew to evade. Less than three hours later, Canberras hit Almaza again and for the first time met light anti-aircraft fire while the lights of the neighbouring Egyptian capital were finally switched off. While frequently assessed as 'ineffectual' – and certainly enough there is no doubt that up to 50 percent of British bombs fell outside the perimeters of their targets – at least one aspect of this offensive proved highly successful: at Almaza, they destroyed 15 out of 20 Curtis C-46 Commando transports of No. 7 Squadron, EAF, which were about to deploy the Egyptian Parachute Brigade into a counter-drop on the Mitla Defile. The mission – which was actually unnecessary as the Israeli paras had withdrawn from the Heitan Defile hours earlier – thus had to be cancelled.[23] Nevertheless, midway through this air strike, the EAF launched its final offensive sorties of the Suez War, in the form of two Il-28s tasked with bombing Tel Nov AB in Israel. The two bombers were piloted by Wing Commander Hamid Abdel-Ghafar and Mustafa Hilmi. Taking off amid British air strikes, Hilmi's aircraft crashed on take-off, killing the entire crew, leaving Abdel-Ghafar to continue the mission alone. While there is little doubt that the low-flying Ilyushin took the Israelis by surprise, any results of this strike remain unknown. The Egyptian then returned to Cairo West only to find the place under another attack by RAF bombers: he landed in complete darkness and while British bombs were going off around him.

Early in the morning of 1 November, the EAF scored its final success of this war, when a pair of MiG-17Fs from No. 1 Squadron intercepted and badly damaged a Canberra PR.Mk 7 over the Nile Delta. The British aircraft returned for a successful emergency landing in Cyprus, but this made it clear that the EAF interceptors were still not neutralised. The Egyptians launched additional air strikes on Israeli ground troops in Sinai shortly after dawn: at the time the Army was still holding out in the Abu Aweigla area, and Amer hoped to it could hold the line from Lake Bardawil via the Gidi Pass to the Mitla Defile. While in a perfect position to repeat its performance from the previous day, the EAF was not given a chance: immediately afterwards, its air bases were subjected to a seemingly endless stream of air strikes flown by British and French fighter-bombers from Cyprus and aircraft carriers in the Mediterranean Sea, which primarily sought to knock out aircraft with unguided rockets and gun fire. Abu Suweir, Almaza and Kabrit found themselves on the receiving end of particularly heavy attacks, but ultimately, it was Abu Suweir – acting as a home for 41 MiG-15bis destined for Syria, half of which were still disassembled – that attracted most Anglo-French attention. Eventually, up to 20 assembled MiGs were destroyed on the ground, and all the disassembled examples damaged.[24] At Almaza, air strikes knocked out at least one Il-28, while another was claimed as shot down by French F-84Fs based in Israel.[25] Up to six MiG-17Fs that were flown in combat earlier that morning were knocked out while still parked on the main apron, followed by eight that were covered by camouflage nets. Four surviving MiG-17s were then flown out by Egyptian pilots. At Fayid, eight Meteor F.Mk 8s of No. 5 Squadron were destroyed on the ground, while the unit's sole surviving Meteor T.Mk 7 was evacuated to Cairo International, where it survived the war.[26] Cairo West, Kabrit and Dikhelia all received similar treatment, but saw the loss of very few MiG-15s because most had now been evacuated. Overall, the British and French fighter-bomber pilots claimed the destruction of 82 MiG-15s and 72 other aircraft on 1 November 1956, together with the probable destruction of 20 MiGs and nine other aircraft, and damage to 41 MiGs and 85 other aircraft.[27]

Obviously, this was on the enthusiastic side as the EAF simply had not had this many aircraft on the attacked air bases for at least 24 hours. However, there is no doubt that the Egyptian Air Force was heavily hit, and lost lots of equipment. Moreover, sustained air strikes had a huge impact upon the Egyptian officer corps, whose officers had previously had social ties to the RAF: they now felt outright betrayed by those they had regarded as friends and appalled at having to watch the enemy flying overhead while they themselves wondered why something was not done to stop it.

NASSER'S DECISIONS

As their air raids continued with increasing intensity through the day, the British and French took control of Egypt's skies. The fact that the EAF did not attempt to intercept them was, at that time, attributed to low morale or even cowardice, and eventually lead to the conclusion that the Egyptian Air Force had been completely destroyed.[28] Only later did it become known that Egyptian pilots and ground personnel wanted to fight back, but were specifically ordered not to fly – at least not for the time being – by President Nasser. In fact, after touring the streets of Cairo to judge the mood of the people during the evening of 31 October and knowing that Egypt's armed forces could not defeat three powerful invaders in open battle, he devised an entirely new grand strategy for his nation's armed forces. Emphasising guerrilla-style resistance in the event that the British and the French would launch an advance on Cairo with the aim of toppling his government and destroying the Egyptian armed services, Nasser ordered not only a complete and immediate withdrawal from Sinai, but also a complete dispersal of the EAF. While most of his military commanders disagreed with their president, and Egyptian commanders in charge of units staffed by Palestinians in the Gaza Strip flatly refused to follow his orders, Nasser remained insistent. Left without a choice, Amer thus forwarded corresponding orders to all units in the field. The result was two-fold: the EAF disappeared from the skies, while in Sinai the Egyptian Army abandoned its positions in a hurry and – left without air cover – left much of its brand-new equipment behind. It is for this reason that for the following five days, the officers and other ranks of the Egyptian Air Force essentially stood idle, while their air bases and equipment were savaged by one British and French air strike after another. Indeed, the dispersal of the EAF reached such proportions that before long the aggressors launched a hunt for its remaining aircraft. On 2 November, French Air Force units deployed in Israel raided Luxor Airport, claiming the destruction six Il-28 bombers. Ultimately, the British and French aircrews claimed the destruction of a further 28 MiGs and 28 other aircraft, the probable destruction of two MiGs and six other aircraft, and damage to 32 MiGs and 75 other aircraft during the day.[29]

Even this proved unsatisfactory for their commanders. Reports about a 'significant concentration of MiGs' at Almaza AB, provoked a series of devastating air strikes on that facility early on 3 November. Moreover, and supposedly concerned about a threat of air attacks on their 'exposed' air bases on Cyprus, Anglo-French commanders began ordering intensive aerial reconnaissance of Syria, resulting in reports – among others – about the sighting of four Il-28 bombers at al-Mezzeh AB outside Damascus on 2 November.[30] Ultimately, the British and French claimed the destruction of another four MiGs and 17 other aircraft, plus damage to three MiGs and 22 other aircraft in Egypt on 3 November 1956.[31]

On the morning of 4 November, the French deployed in Israel launched another raid on Luxor Airport. This time, 13 F-84Fs claimed the destruction 20 Il-28s on the ground. Around noon, six additional Thunderjets repeated this attack, followed by a single RF-84F from Cyprus, which took post-strike reconnaissance photographs. Eventually, the AdA claimed the destruction of 17 EAF aircraft on that day, including at least 10 Il-28s. For comparison, Egyptian sources confirmed the loss of six Il-28s, but also that at least one bomber survived this onslaught and was subsequently flown to the safety of Jeddah airport in Saudi Arabia. Eventually, finding ever fewer Egyptian aircraft to strike, the British and the French then began striking EAF radar stations and storage facilities, before, on 5 November, concentrating on Egyptian Army positions in areas about to be subjected to airborne and amphibious assaults. However, by then it became known that London and Paris were being subjected to increasing international pressure to stop their aggression and withdraw. Indeed, outraged by the Anglo-French intervention and knowing the US public in general were strictly opposed to such actions, President Eisenhower sponsored a UN resolution denouncing the tripartite aggression on Egypt. Issued on 2 November 1956, this called for an immediate ceasefire. When Great Britain, France and Israel failed to stop, Eisenhower threatened to cancel vital loans to them. In an attempt to create a fait accompli, the Anglo-French forces then launched a hurried invasion of Port Said – now vacated by most regular Egyptian units – and then pushed down both sides of the Suez Canal, while the Israelis accelerated their advance into Sinai. However, it was now too late: ultimately, both London and Paris were forced to accept a ceasefire on 6 November 1956. Ironically, it was at dawn of that day that the British troops in control of the disused Gamil airfield, a few kilometres west of Port Said, found themselves exposed to an attack by two MiG-15bis of the EAF. In an obvious demonstration of defiance, the jets launched from a stretch of straight road near Benha, led by Flight Officer Nabil Kamil from No. 1 Squadron and flew this attack to show that the Egyptian Air Force still existed. Rather unsurprisingly considering enemy superiority, one of the two jets returned to Benha with 32 small-calibre holes in its fuselage. Certainly enough, this re-appearance of the EAF prompted additional air strikes on its bases in which the British and French pilots claimed the destruction of two further MiGs, and five other aircraft, the probable destruction of one other aircraft, and damage to five MiGs and 16 other aircraft. The war was – for all practical purposes – over, at least in Egypt.

THE FINAL TALLY

From the Egyptian perspective, the Suez War went on for 120 days longer, until the Anglo-French forces withdrew. Indeed, it was over only once another intervention by President Eisenhower forced Israel to withdraw from the Sinai Peninsula in March 1957 – though only once the IDF summarily executed dozens of Palestinian troops and their Egyptian officers, as well as between 930 and 1,200 other people in Gaza.[32]

Because Egypt never published an official list of its losses during the Suez War, the British and French claims for the destruction of 124 MiG-15s and MiG-17s between 1 and 6 November 1956 are widely accepted as solid, and thus the EAF considered to have been completely wiped out.[33] Something entirely different should have been realised by all the informed observers but was not. In mid-November 1956 when, upon the return of most of the Czechoslovak advisors from their temporary exile in Sudan, and urgent repairs to its major air bases – the EAF also began returning its evacuated aircraft, usually in full view of the British and French, and definitely in front of various UN representatives. By December 1956, a total of 32 MiG-15bis and six MiG-17Fs were either back in Egypt, or repaired from aircraft damaged on the ground during the Suez War. Furthermore, while 21 MiG-15bis destined for Syria and already assembled were destroyed, the 20 non-assembled examples survived the conflict and – with permission from Damascus – were subsequently integrated into the EAF.[34]

The number of Il-28s which survived the Suez War is much more difficult to assess. Egyptian sources confirm the loss of eight bombers between 1 and 6 November. However, apart from the evacuation of about 20 Il-28s by Czechoslovak and Soviet instructors, and by Usama Sidqi, to Jeddah in Saudi Arabia, the whereabouts of the remaining 17 aircraft of this type remain unclear. One way or the other, the conclusion is that at least 40 percent of the aircraft ordered from Czechoslovakia and the USSR in 1955-1956 survived the Suez War. That the actual figure was probably higher became obvious when the Egyptians and Czechoslovaks concluded their third arms deal, on 28 February 1957, stipulating orders for the aircraft listed in the Table 3. Notably, by then the influence of the Czechoslovak advisors – who not only organised a lessons learned session for the top brass of the EAF, but also helped reconstruct damaged air bases – resulted in the decision to bolster and upgrade the Egyptian fighter-fleet with much improved MiG-17s. On the other hand, the Third Czech Arms Deal made the extent of the almost-complete destruction of the Egyptian training and transport fleets obvious: indeed, it prompted the EAF into placing orders for more training aircraft than it had before the Suez War.

Table 3: List of Aircraft ordered for the EAF with the 3rd Czechoslovak Arms Deal, 28 February 1957[35]

Aircraft Type	Number
MiG-15bis	30
MiG-15UTI	6
MiG-17F	77
Il-28	20 newly built, 10 second-hand aircraft
Il-28U	2
Yak-11	40
Zlin Z.226T	50
L-60	10
Il-14	5
Mi-1	5
Mi-4	5

IGNORED LESSONS

The Egyptian armed forces should have learned a number of important lessons from the Suez War of 1956, but failed to do so. Instead, especially the EAF subsequently began showing weaknesses that were to remain typical in its methods of operations for the next decade. Although proving able to operate brand-new equipment entirely on its own, the Air Force found itself on the defensive right from the start of the conflict and, with fewer than a handful of exceptions, never arrived at the idea of operating offensively against Israel. Certainly enough, part of the reason was that its offensive capabilities were underdeveloped: the Il-28 fleet was not ready at the start of the war. Nevertheless, it seems that nobody in Egypt ran a closer study of their few raids against Hatzor and Tel Nov: indeed, even though everybody was angry about what happened in 1956, the prevalent feeling was that Egypt was betrayed by Great Britain and should grow strong to resist such an aggression if it should happen again – but it seems that nobody sought the capability to hit back.

Similarly, the fact that the mass of EAF aircraft were lost because of the failure to fortify air bases was completely ignored. Likewise, instead of studying the weaknesses of the MiG-15s and MiG-17s in comparison to modern fighter-bombers of French origin, as operated by the IDF/AF, the Egyptians returned to the practice of buying new equipment and training ever more pilots, without paying attention to improving the available aircraft or even adapting them to the local conditions. While some of this can be explained by the conclusion that the majority of Egyptian pilots were inexperienced novices, there is no doubt that their superiors failed to teach them about

HAFEZ AL-ASSAD: METEOR PILOT

By accident or design, the late 1950s and the 1960s were fateful times – for many MiG pilots and for most of the Arab states. Ironically, there is hardly a better example of this than the story of Hafez al-Assad, the ruler of Syria from 1970 until 2000, and father of the current leader in Damascus, Bashar al-Assad. Amongst various myths still making circles today, Assad was one of the first 15 Syrian military pilots, enjoyed a stellar career with the flying service, and his subsequent ascent to power was directly related to him – personally – building-up the Syrian Arab Air Force. Beyond that, details are very scarce – if provided at all. As is the case with so many other legends, most of such stories are well-distanced from the truth.

In reality, and as described in Chapter 1, the first group of SyAAF pilots and ground personnel was trained by a group of seven Croat and German instructors to fly US-made T-6 Texans and Piper Cubs, thus creating the core of the Air Force that took part into the Palestine War. Certainly enough, Hafez al-Assad was not amongst them. He joined the Military Academy in Homs only in 1950. Four years later, upon hearing that the SyAAF was in the process of establishing its Air Force College at Nayrab AB, he requested reassignment there. Under the tuition of instructors like Captain Fouad Kallas (who later rose to command the SyAAF), he learned to fly Chipmunks, Texans and Fiat G.46 and G.55 fighters, before graduating in 1955.

Meanwhile, in 1952-1953, Syria had acquired Meteor F.Mk 9s. Because the SyAAF lacked advanced training facilities, its future jet pilots were all trained abroad, usually in Great Britain or Egypt, and sometimes in Italy. As one of the top students in his class, Assad was selected for a jet conversion course in Great Britain. However, this was cancelled in the light of another London-imposed arms embargo, and thus the young Syrian underwent a six-month course at the Air Force College in Bilbeis, where he was instructed by – amongst others – Hosni Mubarak. Back in Syria, Hafez was assigned to No. 9 Squadron, commanded by Squadron Leader Munir al-Garudy, and soon became involved in efforts to intercept British, French and US reconnaissance aircraft that regularly overflew Lebanon, Syria and Iraq in the mid-1950s. At the time, the SyAAF had no early warning radars, but one of its fighter controllers and Tahsin Zaki, Egyptian Air Force Attaché in Damascus, developed a plan to catch an intruder with the help of information provided by local police stations via telephone.

While failing to catch any intruders on earlier occasions, the first time the Syrians managed to bring one of their interceptors into the air in time was on the morning of 6 November 1956, when Assad was scrambled from Nayrab to intercept a lonesome intruder reported by one of the police stations in the Latakia area. Although he managed to approach his target close enough to open fire, Assad's shells missed and the Canberra escaped in the direction of Cyprus. Only hours later, two Meteors led by Garudy were scrambled from al-Mezzeh AB, and two – including one flown by Assad – from Nayrab. Once again, their target was a RAF Canberra announced by a telephone call from the Latakia area. This time, the Syrian interceptors caught up with the British aircraft somewhere over the Homs area and attacked it from two sides. Although the Canberra pilot made a hard turn and attempted to escape in the direction of Lebanon, it was too late: shortly before crossing the border the bomber received a hit in its starboard engine and two of the crew ejected: the third crewmember failed to bail out and was killed when the doomed aircraft crashed – only metres inside Syria.

That was not the end of the excitement for the SyAAF that day. Shortly before sunset, Assad was scrambled from Nayrab AB for the third time, apparently in reaction to the appearance of a lonesome Lockheed U-2 of the CIA that – just like the British before – was also searching for famed Soviet MiGs. Certainly enough, there was no way Assad could reach his high-flying target in a Meteor. Disappointed, he returned to land at Nayrab after sunset. Due to a malfunction of his radio, he could not receive a warning from the control tower, advising him that the wind had changed direction. Unsurprisingly – and because he scrambled even though his brakes had failed before take-off – Hafez al-Assad overshot on landing and crashed his jet against a low wall made of rocks. Thrown out of his cockpit by the force of impact, he was lucky enough to land atop a tent in the middle of a nearby Palestinian refugee camp, with only minor injuries. Not the least pleased about the loss of a precious Meteor, Garudy reprimanded Assad and issued him a – suspended – jail sentence.

Certainly enough, overflights of Syria continued for years longer: as far as is known, none ever found any Soviet MiGs before the first MiG-17s for the SyAAF were actually delivered. Nevertheless, for years after, the version circulated in the Western press was that the RAF Canberra shot down on 6 November 1956 was intercepted by 'Soviet pilots flying Syrian MiGs'.

A group of young Syrian pilots at the Air Force College at Nayrab Airport in 1955. Leaning on the cockpit of the Fiat G.59/2b in the background is Hafez al-Assad. (David Nicolle Collection)

A pair of Meteor F.Mk 9s of the SyAAF seen prior to their delivery to Syria in 1952-1953. Essentially obsolete by the time, they still represented a quantum leap forward over the earlier, piston-engined fighters operated by the SyAAF. (Gloster, via Tom Cooper)

Shalaby el-Hinnawy inspecting a brand-new MiG-17F (construction number 8047), delivered after the Suez War. His warnings about ignoring the lessons of that conflict were lost in the overall euphoria over the political victory. (David Nicolle Collection)

A smaller roundel applied on the rear fuselage identifies the MiG-17F (without serial) in this beautiful study as one of the examples delivered to Egypt after the Suez War. Notable are identifications stripes applied around the wingtips and on the rear fuselage. (David Nicolle Collection)

new lessons. This was mainly to be seen at the tactical level, where EAF MiG pilots began using staggered formations, as instructed by the Czechoslovaks and Soviets, and – when under attack – tried to evade with climbing turns. Their aircraft actually had poorer climbing performance than the Mystères, and they lacked power-assisted controls to lessen stress on the pilot during hard turns. Unsurprisingly, the Egyptian pilots disliked such manoeuvres, but never developed their own.

That all said, the conclusion is unavoidable that the principal problem was something the Egyptians could do very little about: the core design of the Soviet-made equipment and armament. MiG-15s and MiG-17s were optimised for high altitude operations and intercepts of heavy bombers, like the US-made Boeing B-29. Whenever operating at low altitudes, their engines proved true gas-guzzlers, greatly limiting their effective range. This was slightly improved through the installation of drop tanks, but the MiG-15bis arrived still equipped with slipper-type tanks, attached directly to the wing: if one failed to separate when jettisoned – as was usually done at the start of an air combat – the aircraft was almost certain to fall into a spin. Due to intensity of operations on 30 October 1956, drop tanks were rapidly expended: by the second day of operations, there was already a shortage of them.

Because of their optimisation for intercepts of high-flying bombers, both MiGs came armed with two Nudleman-Rikter NR-23 and one Nudleman N-37D gun, packed – together with their ammunition – inside a practical container under the nose of the jet. The former had a calibre of 23mm with a rate of fire of 850 to 950 rounds per minute, but very low muzzle velocity (just 690 metres per second, or 2,264ft/sec); the later had the same muzzle velocity, but a rate of fire of only 400 rounds per minute. In other words: Soviet guns fired heavy shells (200g for 23mm and 700g for 37mm), but at a low rate. Moreover, the shells fired from the NR-23 and the N-37 flew along entirely different trajectories. While the Soviets estimated that two hits by 37mm rounds, or eight by 23mm, should be sufficient to shoot down any bomber, in reality, the different flight trajectory of their shells resulted in significant sighting issues, which meant that they rarely hit. Moreover, because both the MiG-15bis and MiG-17F packed only 40 rounds of 37mm calibre, and 80 for each of the two NR-23s, they were always critically short of ammunition, and unlikely to be able to repeat their attacks in the event of a miss. Fighters armed this way were at a distinct disadvantage vis-à-vis Western designs. The Mystère IVA was equipped with two 30mm DEFA 551 cannons (a further development of the German Mk.213/30 from the Second World War), which had a higher muzzle velocity (800m/sec or 2,625ft/sec), and an almost three times higher rate of fire. The four Hispano HS.404 cannons installed on Ouragans had very similar performance – and, like the Mystère, this jet had power-assisted flight controls, a much better gunsight and carried 150 rounds per gun. Finally, even if lighter, the shells fired from DEFAs and HS.404s proved far more reliable and possessed higher destructive power than Czechoslovak and

While MiG-15bis were the first combat aircraft of Soviet origin acquired by Egypt and Syria in 1955, the way to maintain and operate them was led by MiG-15UTI two-seat conversion trainers. The construction of these was the same as that of the single-seat interceptor, but they included a second cockpit (usually occupied by the instructor pilot) instead of the front fuel tank and, instead of two 23mm and one 37mm cannon, had only one 12.7mm machine gun installed in the lower forward fuselage. Regardless of whether single or two-seaters, all the MiG-15s were originally painted in the factory in two layers of clear lacquer mixed with 10 percent and 5 percent aluminium powder, respectively, applied directly over metal surfaces. This gave them the overall appearance colloquially called 'silver grey'. Serials of the first batch of Egyptian MiG-15UTIs were in the range 1406-1448. (Artwork by Tom Cooper)

A reconstruction of a typical MiG-15bis as assigned to Nos. 1 and 30 Squadrons in 1955 and during the Suez War of 1956. They had no rear-view mirrors atop the cockpit transparency, wore much bigger roundels (applied in six positions) than later on, had two black identification stripes around the rear fuselage (often widely separated), and – in the same fashion as earlier Vampires and Meteors – three identification stripes around the wingtips (the centre one was at least twice as thick as the two outside stripes). Such markings remained in widespread use until 1958 when UARAF roundels and fin flashes were applied. Although there is next to no photographic evidence, pilot logbooks of Firky el-Gahramy, Fikry el-Gindy and Abdel Moneim el-Shennawy show serials in the 1900-1997 range, applied non-sequentially, for example: 1910, 1950, 1956, 1968, 1972, 1975 and as shown here 1977. (Artwork by Tom Cooper)

A reconstruction of perhaps the MiG-15bis of the Suez War best-known in the West: the jet piloted by Abd ar-Rahman Muharram on 31 October 1956, that ditched in Lake Bardawil after running out of fuel due to combat damage and was subsequently recovered by the Israelis. Photographs of it still wearing the serial number (this was removed soon after the aircraft was brought to Israel) are rare and not entirely clear. Its serial is known to have ended with 01, leading to the conclusion that the complete number was 1901. All the early Egyptian MiG-15s used slipper-type drop tanks only, which Muharram jettisoned before his belly landing: these were replaced by those taken from MiG-17s only in the mid-1960s. (Artwork by Tom Cooper)

The first batch of MiG-17Fs, delivered to Egypt in October 1956, wore the same 'silver grey' livery as the MiG-15s and had their identification stripes applied in similar fashion. They had roundels on their rear fuselages applied in same size as on the wings – and were thus much bigger than was subsequently the practice. As far as is known, these aircraft wore serials in the 2000 range, like 2017 shown here, and also 2004, 2028, 2034, 2060, 2072, 2082 and 2084. Most had seen intensive service in a number of crises and wars involving Egypt in the 1950s and the 1960s and were gradually modified through the addition of launch rails for unguided rockets and then through the addition of hardpoints capable of carrying 250kg under the lower fuselage. This example is known to have survived not only the Suez War of 1956, but the war in Yemen of 1962-1967 and the June 1967 Arab-Israeli War (which are to be covered in the Volume 2). (Artwork by Tom Cooper)

In 1957, the Fighter Training Unit of the EAF began applying this insignia on its MiG-15bis and MiG-15UTIs (also on this example, serial number 1968, an obvious survivor from the original batch of 80 MiGs delivered in 1955). Apparently inspired by ancient Egyptian mythology, this crest showed an Egyptian mythical creature in red and yellow, with blue 'wings' on either side. As visible from the inset photograph, as well as for training Egyptian personnel, jets of the FTU were also used to train a team of Indonesian pilots and ground technicians. Notably, all maintenance markings on these early Egyptian MiG-15s (and MiG-17Fs) were applied in enamel blue in English, while four-digit construction numbers were applied in red on all movable or removable parts of the aircraft, including various covers, flaps and the rudder. (Artwork by Tom Cooper; Photo via Marc Koelich)

Hard on the heels of the FTU, either the re-established No. 5 Squadron or No. 20 Squadron followed in fashion and began applying unit insignia on the forward fuselages of its MiG-15UTIs and MiG-15bis. Apparently based on Coptic mythology (the George and the Dragon legend), this was reportedly inherited from No. 20 Squadron from the times when this unit flew Vampire FB.Mk 52s, and consisted of a black serpent, applied in one of two slightly different ways, as shown in the insets. The MiG-15UTI shown here, serial number 1998, belonged to the second batch of such aircraft: the six examples wearing serials in the range 1993-1998, and acquired as a part of the Third Czechoslovak Arms Deal, in 1957. They were the first Egyptian MiG-15s to be equipped with MiG-17-type of drop tanks. MiG-15UTI serial number 1998 is known to have retained this insignia through the duration of the UAR and well into the 1960s. (Artwork by Tom Cooper)

Another survivor of the Suez War 1956 was the MiG-15bis serial number 1990. Originally assigned to either No. 1 or No. 30 Squadron, in 1957 it was re-assigned to No. 5 or No. 20 Squadron – while this unit was training a group of Indonesian pilots on this type – and thus received the 'Black Serpent' insignia while still wearing the EAF national markings, including large roundels on the rear fuselage. This MiG-15bis retained the same insignia once it received the UARAF insignia – in red, white and black, with two green stars on the white field – in February 1958 and was deployed for several tours of duty to el-Arish AB during tensions with Israel in 1958-1961. In typical Egyptian fashion, because of its unit insignia by then No. 5 Squadron was colloquially known in the Air Force as 'The Worms'. (Artwork by Tom Cooper; photo by Fikry el-Gahramy)

Not only operational units, but also the FTU of the EAF regularly deployed temporary detachments of its MiG-15bis to el-Arish AB during the tensions with Israel in 1958-1961. Amongst others, this is confirmed by the fact that on 8 January 1959, Abdel Moneim el-Shennawy claimed a Mystère IVA of the IDF/AF shot down over the Sinai Peninsula while flying this MiG-15bis, serial number 1956 (and thus a survivor of the Suez War in 1956). The pilot was still assigned to the FTU and his aircraft wore that unit's insignia. By the time of the 1959 incident the jet was marked with the full UARAF insignia, including roundels applied in six positions, and the new fin flash. Identification stripes of the EAF were retained in their usual positions, as were maintenance stencils in enamel blue and the last four of the construction number in red. (Artwork by Tom Cooper)

The third EAF/UARAF unit known to have introduced the practice of applying unit insignia on its aircraft was No. 30 Squadron, for which Tahsin Zaki designed this 'Leopard's Head' in black and white, applied on a red disc, and completed with wings and a streak in yellow outlined in black. It is possible that this eventually led to the development of the famous 'Flying Leopard' insignia, applied on many MiG-15s and MiG-17s of No. 18 Squadron as of June 1967 (to be discussed in the Volume 2). Abdel Moneim el-Shennawy flew this jet in his first air combat with the Israelis, on 11 December 1958. This illustration shows the jet about one or two years later, after the addition of two double launch rails for Sakr unguided rockets (designed by Hispano-Suisa and manufactured under licence in Egypt since at least 1952) under each wingtip. First installed on Egyptian MiG-15bis shortly before the Suez War of 1956, Sakrs were available in two versions: one with an armour-piercing warhead and the other with an anti-personnel/soft-target warhead. (Artwork by Tom Cooper; photo via Abdallah Emran)

The appearance of the first 41 MiG-15bis and four MiG-15UTIs of the SyAAF – all acquired from Czechoslovakia – remains unknown except that the single-seaters wore serials in the range 900-999 (which the EAF retained after Damascus donated survivors of the Suez War to Cairo, in 1956). This is one of the MiG-17Fs acquired directly from Moscow in 1957 and was probably assigned to No. 9 Squadron by the time it was inspected by Taher Zaki in his function as the EAF attaché later the same year. Essentially, the livery was the same silver grey as usual, and maintenance stencils should have been applied in English, in enamel blue. The SyAAF applied serial numbers on its aircraft in a significantly different way than the EAF: the full serial – 946 in this case – near the top of the fin, while the 'last two' were 'repeated' on the forward fuselage, almost in the style of the Soviet Air Force, but in matt black Persian digits. (Artwork by Tom Cooper)

Out of 45 Il-28s delivered to Egypt in late 1955 and early 1956, only some 12 were regularly operated by the time of the Suez War, and the example coded as 'K' was one of them. Like the rest of the fleet, the jet arrived painted in the same 'silver grey' livery as the MiGs, which is in two layers of clear lacquer, mixed with 10 percent and 5 percent aluminium powder, respectively. Roundels were applied in six positions, with the ones on either side of the rear fuselage relatively small and directly below the fin. The code was repeated – in small black letters and digits – on the cover of the front undercarriage bay, and the cover for entry into the gunner's position below the rear fuselage. Insets in the right lower corner show the principal armament of Egyptian Il-28s, including (from left to right) FAB-250M-46, FAB-100M-46 and FAB-50M-46 bombs. (Artwork by Tom Cooper)

Within the framework of the Third Czechoslovak Arms Deal, Egypt ordered 30 additional Il-28s (including 20 newly built and 10 second-hand aircraft) and two Il-28Us. These were delivered in time to enter service before the creation of the UARAF, and received the same set of national markings. The roundels applied on their rear fuselages were significantly larger than those on the original Il-28s, and positioned further forward. Like examples delivered under the First Czechoslovak Arms Deal, they received single-letter codes: on some of aircraft (like C1, shown here, or V1), these were suffixed by '1': this identified new aircraft, replacing those destroyed during the Suez War of 1956. (Artwork by Tom Cooper)

Soon after the establishment of the UARAF all of the brand-new Syrian MiG-17PFs were transferred to Egypt. Once there they were assigned to No. 31 Squadron – the EAF's first night-fighter unit, which previously operated Meteor NF.Mk 13s. MiG-17PFs retained their silver grey overall livery, but had the radar cover for the upper antenna painted in blue-green: the cover for the lower antenna (inside the intake) was sometimes painted in the same colour but at least as often in white, as shown here. Other markings included serials in the range 2800-2899 (known examples were 2802, 2803 and 2816 shown here). The fourth EAF/UARAF unit known to have applied crests on its aircraft, No. 31 Squadron selected this 'Crow-Bat atop a Globe' in black, with a backdrop consisting of a white circle with black outline: it remained in use at least until the June 1967 War with Israel. (Artwork by Tom Cooper; photo from Nour Bardai Collection)

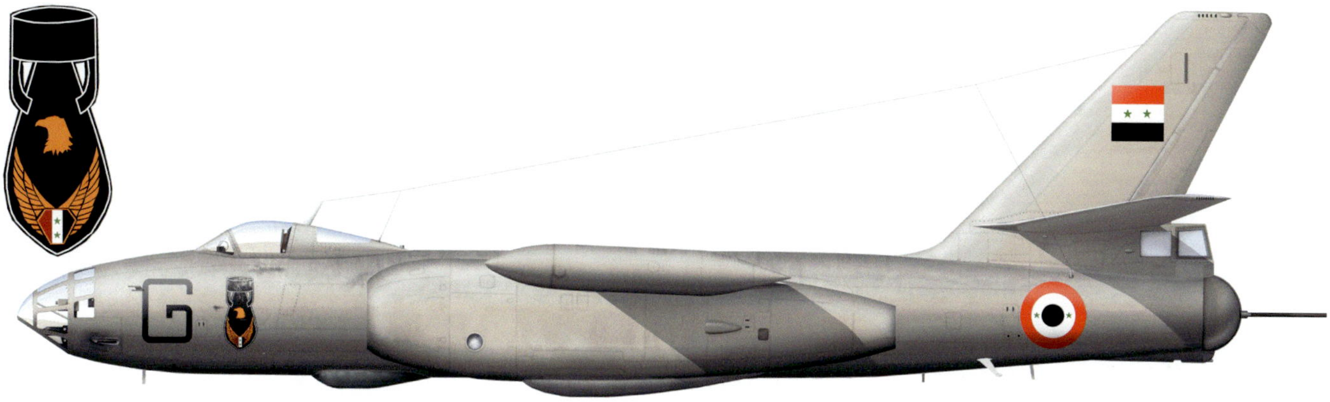

Il-28-equipped squadrons of Air Group 61 began applying unit insignia on their aircraft in 1958. This consisted of a large bomb in black and white, applied by hand with a brush. Inside this was a shield with the UAR flag, surrounded by yellow wings, above which the individual squadron mascot was applied in yellow: while it remains unknown which was used by Nos. 8 and 9 Squadrons, this was either an 'Eagle Head', or 'Cobra Head'. While it also remains unknown if all Il-28s received such markings, several Il-28Rs involved in nocturnal operations over Israel did. This was the case with the Il-28R coded 'G', illustrated here, which was involved in the overflight on 21 September 1961. Notably, gauging by the position of the roundel on the rear of its fuselage, and the lack of suffix '1' in its code, this jet probably belonged to the original, first batch of 48 Il-28s acquired during the First Czechoslovak Arms Deal. (Artwork by Tom Cooper)

All the 56 MiG-19Ss introduced to service with the UARAF in 1961-1962 were painted in the Soviet colour designated AS-1115 (also MRP-354), later known as 'Russian Air Superiority Grey'. Applied overall, this was a satin mid-grey colour which tended to develop a light bluish touch due to wear. Their serials were in the 3000-3999 range. Recently discovered photographs have confirmed the application of 'traditional' identification stripes around the rear fuselage and wingtips: these were always applied in black, without any kind of white fields in between. Due to the poor manufacturing quality and mechanical unreliability, the type suffered heavy attrition, further increased by its involvement in the Yemen War and the June 1967 Arab-Israeli War. The example illustrated here – serial number 3024 – in UARAF markings of early 1961, survived this turbulent period and is presently on display at the EAF Museum in Almaza. (Artwork by Tom Cooper; photo by Tarek el-Shennawy)

As in the case of the EAF and the SyAAF, some of the first combat aircraft of Soviet origin acquired by the IrAF were MiG-15UTIs. Although at least one was usually assigned to units operating MiG-17Fs, the majority of Iraqi MiG-15UTIs served as advanced trainers at the Air Force College, where instructors seconded from the Indian Air Force used them to convert Iraqi pilots to faster MiGs. As usual, they retained their 'silver grey' overall livery: national markings were applied in six positions, and fin flashes on the fin, while the serial was frequently repeated on the drop tanks – all of which were of MiG-17-type. (Artwork by Tom Cooper)

Because of long-standing Iraqi designs to establish three bomber units – and the British refusal to deliver any English Electric Canberras – the next Soviet-made type to enter service with IrAF was the Il-28. All the 24 examples known to have been acquired between 1958 and 1963 were operated by No. 8 Squadron (the Tiger-containing crest of which is shown inset). While retaining their overall 'silver grey' livery, they wore the national insignia in six positions. This example is shown with the initial version of the fin flash as used following the 14 Tammuz Revolution. This consisted of the old four-colour fin flash, now decorated with a yellow 'sun' in the centre, subsequently replaced by a three-colour fin flash in green, white and black, with a yellow and red 'sun' on the white field. Iraqi Il-28s saw much action against Kurdish insurgents led by Barzani all through the 1960s and well into the 1970s. (Artwork by Tom Cooper)

By the time No. 5 Squadron, IrAF, became operational on MiG-17Fs in March 1959, they had received the new fin flash, in green, white and black, with a yellow and red 'sun' on a white field: initially applied on the upper part of the fin, this was subsequently moved lower, as shown here. The Iraqis took care to purchase Soviet-made add-on underwing pylons for UB-16-57 pods for 57mm S-5K unguided rockets. MiG-17Fs armed this way are known to have seen their first combat during the anti-communist uprising in Mosul which erupted on 7 March 1959 and left much of the city in ruin. Notably, Iraqi MiG-17Fs never received any national insignia on their top wing surfaces: only oversized insignia on the undersides, shown in the right inset. The other inset shows the crest of No. 5 Squadron, which dated back to the late 1930s. (Artwork by Tom Cooper)

Iraqi MiG-17PFs wore the same 'silver grey' livery as MiG-15UTIs and MiG-17Fs, and had their radar antennas painted in blue-green and white. They had an IFF antenna atop the fuselage. Applied in sequence of delivery, as on all the Iraqi military aircraft, their serials were in the range 452-469. Although frequently deployed as fighter-bombers, as far as is known they never used add-on underwing pylons for UB-16-57 pods, as widely deployed by MiG-17Fs. Inset is shown the insignia of No. 7 Squadron, a unit as colourful as Iraq's history: originally established in 1939 (when it flew US-made Northrop 8A-4D bombers), the squadron was destroyed in 1941 but then re-established only a few years later to fly Avro Ansons and then Hawker Furies, and did so during the Palestine War. (Artwork by Tom Cooper)

Because it received its first MiG-19s in late 1959, and began operating them in 1960, Iraq was actually the first Arab country to operate supersonic fighter jets. Moreover, these seem to have been some of the last newly manufactured examples – as production of MiG-19Ss ceased in 1959 – and thus were in far better condition than those subsequently delivered to the UARAF. Iraqi MiG-19Ss were painted in the same 'Russian Air Superiority Grey' overall satin mid-grey colour, and had their national markings applied in six positions, plus a fin flash. Known serials were in the range 489 to 494, 498, 500-504, and 518-521 and were always applied on the forward fuselage (and sometimes repeated on drop tanks). Insets show the insignia of No. 9 Squadron and the ORO-57K pod for S-5K unguided rockets: the latter was usually carried on a small pylon installed behind the underwing undercarriage. (Artwork by Tom Cooper)

Thanks to General Awqati's connections to Moscow, Iraq became the first Arab country to operate MiG-21F-13s, and the first to operate interceptors equipped with guided air-to-air missiles (Soviet-made R-3S, ASCC/NATO-codename 'AA-2 Atoll'). A total of 16 were delivered to the newly established No. 11 Squadron by 1962 (the crest of that unit is shown inset). All were painted in two layers of clear lacquer mixed with 10 percent and 5 percent aluminium powder, overall, giving them the same overall 'silver grey' appearance as on earlier MiG-15s and MiG-17s. Like No. 9 Squadron, No. 11 was dedicated to the protection of Qasim's government in Baghdad, and thus found itself exposed to severe air strikes during the coup of February 1963, when it lost at least four aircraft. Later on, its crews were to find themselves exposed to Israeli attention, with murderous consequences. (Artwork by Tom Cooper)

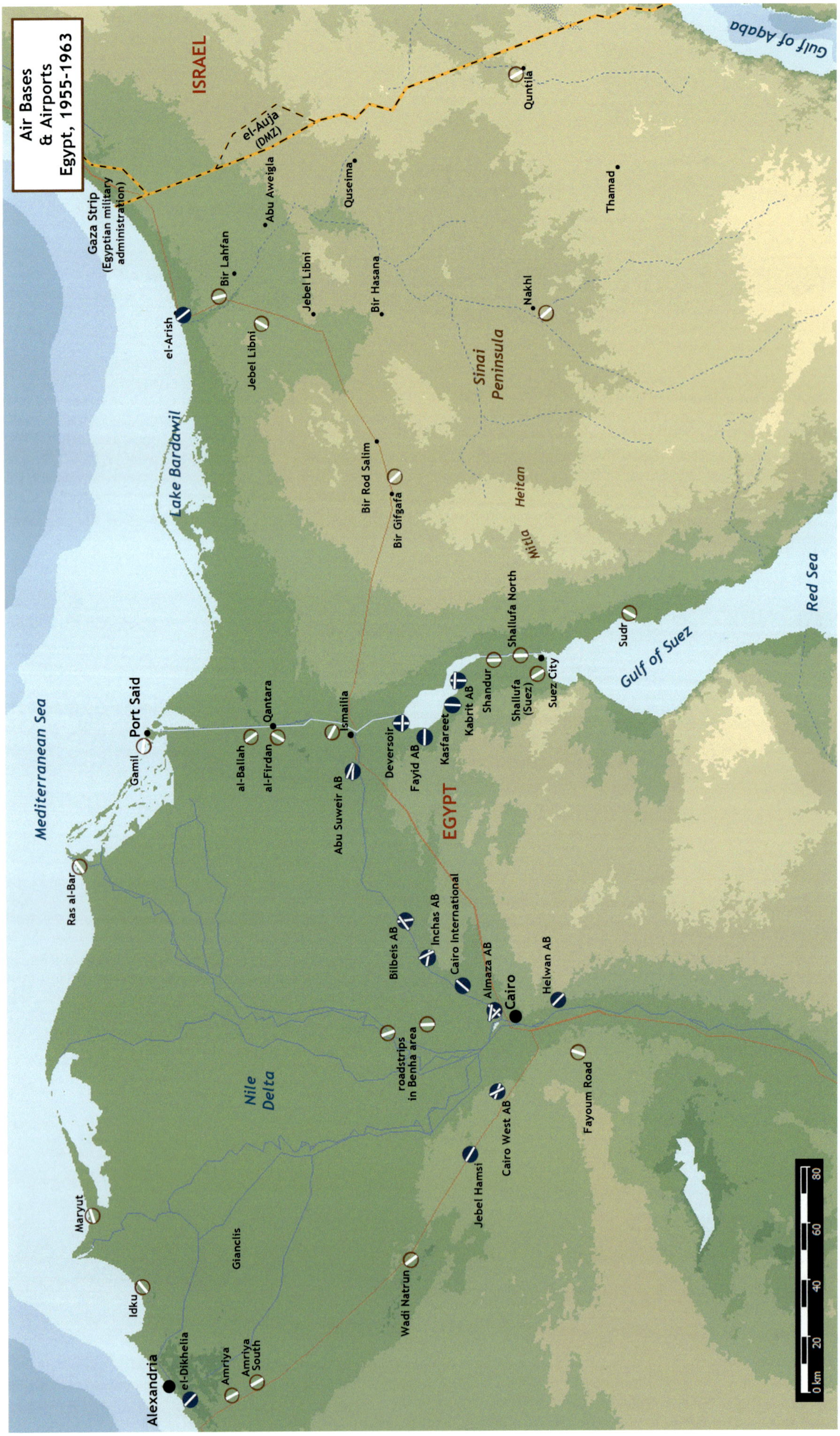

A map of air bases and other airfields in use in Egypt as of 1955-1963. (Map by Tom Cooper)

A UN soldier watching EAF ground crew manhandling a MiG-15bis at Abu Suweir Air Base in the Canal Zone in late 1956. (Albert Grandolini Collection)

Soviet-made shells. As a consequence, although considered sturdy and reliable, MiGs were regularly knocked out of combat by single hits, while Mystères and Ouragans survived multiple hits – which meant that overall, the MiGs were always at a distinct disadvantage in encounters with enemy fighters: their pilots were far less likely to score hits, thus far less likely to knock the enemy out of combat, and thus unlikely to establish aerial superiority. Amazingly enough, the Egyptians, Czechoslovaks and Soviets all failed to realise this fact – with tragic consequences, not only for Egypt, but the entire Arab world.[36]

4

THE GREAT TUMULT

The period following the Suez War of 1956 until the June 1967 Arab-Israeli War is often described as a quiet decade. Certainly enough, after that conflict, the Egyptian and Syrian air forces took time to recover. However, from the Arab point of view, this decade was anything but quiet. On the contrary, tensions with Israel escalated soon enough, this time against the backdrop of the simmering Water War – a massively misreported affair related to the control of several aquifers around Israel, and ongoing since the 1950s. Another issue became the question of Palestinian refugees: people ethnically cleansed out of Palestine during the war of 1947-1949, and then ignored and left to vegetate – often barely a few feet away from the ceasefire lines with the emerging Israel. Last, but not least, a wave of pan-Arabism swept over the Middle East in the aftermath of the Suez War, causing severe unrest, coups d'état, civil wars and further foreign interventions.

REBUILDING THE EAF

In Egypt, the Air Force spent most of 1957 in the reconstruction of its air bases and its entire organisation. Once again, it was foremost Czechoslovak – and not Soviet – advisors that supported this massive effort. By deploying four teams of experts, they helped repair numerous Yak-11 training aircraft damaged during the last war, and then trained Egyptian ground crews in undertaking periodic maintenance of additional Yakovlevs that began arriving later that year, and of the new jets; they converted 30 additional pilots to the newly delivered MiG-17Fs, and a team of air defence experts helped rebuild the Fighter Control System. Later in 1957, the first team of instructors deputed from the Indian Air Force joined the Air Force College in Bilbeis, where – over the next decade – they were to provide theoretic and basic flight courses on Yak-11s for at least 10 classes of future EAF pilots. Emboldened by Moscow and Prague's agreements, and free from the requirement to train their pilots from the bottom up, the Egyptians exploited the opportunity to send ever larger groups of fliers and ground crews to Czechoslovakia – mainly for conversion courses to MiG-15s and MiG-17s. At first 50 future MiG pilots, then another 30 MiG pilots, and then 15 Il-28 crews arrived in Prague, followed by future radar operators, until their numbers overwhelmed the local training facilities. In late 1957, the Czechoslovak Air Force found itself short on instructors for its own requirements to a degree where Moscow had to jump in and start providing training personnel instead.[1]

This was still not all. During the same year, Cairo signed a contract for the acquisition of five military repair plants from Czechoslovakia, including one capable of running overhauls of MiGs. Although marred by a shortage of qualified personnel and funding, and thus repeatedly postponed, this project was realised starting in 1962, when a main aircraft facility was established at Almaza AB. Completed in 1965, thanks to dozens of additional specialists trained in Czechoslovakia, by 1970 this factory became capable of undertaking complex engine repairs on MiG-15s, MiG-17s, MiG-19s and Il-14s. Finally, as well as training literally hundreds of pilots, weapons specialists, radio-technicians, engine and airframe specialists, in 1958 top Czechoslovak aerial warfare experts began providing courses for higher echelons of the EAF, including tactics, operational art, logistics, organisation and planning, communication, navigation, bombardment, air-to-air and air-to-ground gunnery. Overall, the EAF was completely reorganised, re-equipped and re-modelled along a mix of Egyptian and Czechoslovak lines and not Soviet, as generally assumed at the time and ever since.

Although Cairo demanded more, Prague proved unable to deliver. On the one hand, under pressure from Moscow, much equipment was sold to Egypt at greatly reduced prices and very favourable conditions, making this entire effort barely profitable for

On 19 December 1957, the EAF staged an open day at Almaza AB, to celebrate its 25th birthday. On display were all the recently acquired aircraft, including (clockwise): Il-28 bomber, Yak-11, two Zlins, two Mi-1 helicopters, a MiG-17F and a MiG-15bis. Notably, large roundels applied on the fuselage of the MiG-17F indicate that this was one of the examples delivered shortly before the Suez War of 1956. (David Nicolle Collection)

One of the busiest squadrons of the EAF in 1957 and 1958, was the Fighter Conversion Unit of the Air Force College: operating a large number of MiG-15bis, MiG-17Fs, and MiG-15UTIs this was the centre at which over 150 Egyptian pilots were prepared for combat operations on Soviet-designed jets. (Nour Bardai Collection)

To bolster recruitment for the EAF, in 1957 the Egyptians organised an aerobatic team. This included seven MiG-17Fs. (Photo by Tahsin Zaki)

Czechoslovakia. Even so, Egypt gradually proved unable to serve the resulting debts, and was forced into postponing their service, which was something that met little sympathy in Prague. On the other hand, and also because of pressure from Moscow, the Czechoslovak arms industry was simultaneously preoccupied with manufacturing equipment for the needs of its own armed forces, while being prevented from developing more advanced weapon systems. Therefore, Prague found itself unable to offer advanced fighter jets, bombers, and air defence systems to Egypt: henceforth, Czechoslovakia would foremost export its know-how and training aids instead.

RE-EQUIPPED SYAAF

If it was to follow the wishes of the SyAAF senior leadership, their air force was to be completely reorganised and re-equipped. With the exception of intercepting RAF aircraft violating its airspace, the SyAAF did not become involved in the fighting during the Suez War. However, its entire complement of MiG-15s was either destroyed or subsequently donated to the EAF and immediately after the war Damascus cancelled the entire order for related spares from Prague. Instead, during the conflict, in October 1956 the Syrian President Shukri al-Kuwatli travelled to Moscow to negotiate a new arms deal, including 60 MiG-17Fs and radar-equipped MiG-17PFs, 12 Il-28s, six Il-14s, 10 helicopters and 20 training aircraft. Correspondingly, two months later a group of 20 Syrian pilots was sent to Czechoslovakia for related conversion courses, 18 to Poland and about a dozen to the USSR. The Soviets were extremely quick to deliver: the first 12 MiG-17Fs reached Syria in January 1957 – well before there were any Syrian personnel to fly them. As a

ABDEL MONEIM EL-SHENNAWY: SURVIVOR

The Western public is likely to recall the case from 14 May 1957, when – all of a sudden – a young Egyptian plot undergoing conversion training in Czechoslovakia, landed a MiG-15bis of the Czechoslovak Air Force at Schwechat International Airport, outside Vienna in Austria. Contrary to what many in the West expected or even hoped for, this was not a defection – nor was it the end of that pilot's career. Quite on the contrary.

The name of the pilot was Abdel Moneim el-Shennawy. He was born on 9 July 1941, in Nefia, a small village in the Nile Delta, three kilometres outside Tanta, as the seventh of eight brothers. His father died early, but his mother raised her children to the best of her abilities, and each received a good education. Amongst others, Abdel's second-oldest son rose in rank to Major-General of the Egyptian Army. Abdel Moneim joined the EAF in July 1951, graduated from the Air Force College in July 1954 after soloing on a Chipmunk, and flew Vampires during the Suez War. Thus, he was anything but a novice pilot by the time he was sent to Czechoslovakia for a conversion course to MiG-15s in 1957. Indeed, he completed that course a year later, and returned to Egypt to serve as instructor pilot at the Air Force College. As well as working in that function until 1966, Shennawy served tours of duty in operational MiG-15 and MiG-17 squadrons and became involved in about a dozen air combats with the Israelis in 1958-1961 alone. Furthermore, he served a tour of duty in Yemen, and commanded Hodeida AB.[2]

In 1966, Abdel Moneim el-Shennawy was selected to command Egypt's first unit equipped with Sukhoi Su-7BMK fighter-bombers. Following a conversion course at Lugovaya in the USSR, he commanded No. 1 Squadron during the June 1967 War with Israel. About a month later, his squadron was expanded into a brigade with three units, but he still personally led his pilots into another dozen combat operations in 1967, and another six in 1968. In early 1969, Shennawy was transferred to the High Command EAF, and assigned responsibility for all advanced and tactical training activities and was then appointed the Inspector General for Fighter-Bomber Units. By the time of the October 1973 Arab-Israeli War, he had the rank of Brigadier-General and was the most senior EAF officer to fly combat sorties.

Young Abdel Moneim el-Shennawy, around the time of the Suez War. (Tarek el-Shennawy collection)

Shennawy served with the EAF for 32 years, retiring only in 1986, by when he had accumulated 8,650 hours on jet aircraft alone (on 58 different aircraft types); flew and fought in five wars, was officially credited with five kills, and survived five ejections from different aircraft. Following his active service, he joined the agricultural aviation. Tragically, while transferring a Cessna 188 from Egypt to Sudan on 14 September 1989, the engine failed: the aircraft was badly damaged on landing in the desert and burst in flames, making it impossible for the injured Abdel Moneim el-Shennawy to escape.

Abdel Moneim el-Shennawy landed a Czechoslovak Air Force MiG-15bis at Schwechat International on 14 May 1957. The reason for this mishap was the failure of the aircraft's electric system and compass. Both the MiG and its pilot were returned to Czechoslovakia about a week later. (Nour Bardai Collection)

A MiG-17F of the SyAAF seen in late 1957. Notably, the SyAAF had the practice of applying the full serial of its aircraft – in this case 939 – atop of the fin: only the 'last two' were repeated on the forward fuselage, always in black. (David Nicolle Collection)

consequence, it was a group of Egyptian pilots – including Farouq el-Ghazzawy – that presented the brand-new MiGs to the Syrian public for the first time, by flying several of them over Damascus during the National Day on 17 April 1957.[3] The remaining 48 MiG-17Fs and MiG-17PFs followed by the end of 1957, together with a group of about 30 Soviet advisors. Combined with pilots and ground crews trained in Czechoslovakia, Egypt, Poland and the Soviet Union, they enabled the contemporary commander of the SyAAF, Major-General Wadiyah Makabri, to re-organise the force as listed in Table 5. The home bases of most of these units remain unknown.

Table 4: List of Aircraft ordered for the SyAAF from Moscow in October 1956

Aircraft Type	Number
MiG-17F/MiG-17PF	60
Il-14	6
Il-28	12
Mi-4	10
Yak-11/Yak-18	20

Table 5: SyAAF, ORBAT, January 1958

Unit	Equipment
No. 1 Squadron	MiG-15UTI, MiG-17F
No. 2 Squadron	MiG-15UTI, MiG-17F
No. 5 Squadron	MiG-17F
No. 9 Squadron	MiG-17PF
Bomber Squadron	working up; Il-28s to follow in 1958
Air Force College	Yak-18, Yak-11, MiG-15UTI

RUSHED UNION

Latent political instability, frequent unrest and growing pressure from within the armed forces, combined with fear of the well-organised Communist Party of Syria, prompted the top political leadership in Damascus to propose a political union with Egypt. Following extensive negotiations – necessary because Nasser originally preferred a confederation – and amid public euphoria, the Egyptian and Syrian presidents then met on 1 February 1958, to announce a merger of the two countries, creating the United Arab Republic (UAR). A logical consequence of this act was a merger of the EAF and the SyAAF into the United Arab Republic Air Force (UARAF).

A close-up photograph of the SyAAF MiG-17F 946, being inspected by a group of Syrian officers and ground crew, and by Taher Zaki (left side, standing in front of the aircraft), the Egyptian Air Force Attaché in Damascus as of 1956-1957, who played an important role in the downing of the RAF Canberra. (Taher Zaki via David Nicolle)

According to the planning in Cairo and Damascus, this was to become a potent force boasting a strength of 30 operational units. Despite all the enthusiasm, the establishment of a true political and economic union was a lofty goal. The principal reason was that the fiercely independent – and 'pluralist' by their political orientation – Syrians found it hard to accept the Egyptian dictatorship, exclusion of the military from political affairs, and a widespread nationalisation of their economic assets. However, with Syrian negotiators insisting on a 'total union' with Egypt, Nasser decided to act quickly: in a matter of days, all the political parties of Syria – except for the Ba'ath – were ordered to dissolve, and the armed forces were deployed for a crackdown against not only the local Communists, but also all opponents of the union. Immediately after introducing a provisional constitution of the UAR, and although appointing an Egyptian and a Syrian as vice-president of each of the two nominal 'provinces' of the union, Nasser appointed Egyptian officials in almost all of the influential positions.

Despite the delivery of MiGs and Ilyushins, the Syrians never had enough time to complete working up their new units, and thus Gloster Meteors still formed the backbone of the SyAAF's interceptor fleet when Syria joined Egypt to form the United Arab Republic. When Egyptian military photographer Osman Mahmoud travelled to the newly declared Northern Province in February 1958, he was thus able to photograph this row including two Meteor NF.Mk 13 (nearest aircraft wore the serial 472) and …

… seven Meteor F.Mk 9s on the tarmac of al-Mezzeh AB. (Photos by Osman Mahmoud, via Nour Bardai)

A MiG-17F – serial number 99 of the former Syrian Arab Air Force – taking off from al-Mezzeh AB, and already showing full UARAF markings, as well as the IFF antenna on the top of the fuselage. (Nour Bardai Collection)

A long row of MiG-17PFs: originally ordered by Syria, these were delivered just in time to be taken over by No. 31 Squadron, UARAF. Originally commanded by Hinnawy, the unit was subsequently taken over by Garudy, who shared a Canberra-kill from November 1956. Notable near the nose of all the aircraft in this photograph is the black 'Crow-Bat' insignia (see colour section for details). (Nour Bardai Collection)

A MiG-17PF in full afterburner about to get airborne. Notable near the intake is the unit insignia of No. 31 Squadron, which sometimes included a white background. More advanced than the Meteor NF.Mk 13s, which also proved complex to maintain without approach to sources of spares in Great Britain, the type served as the principal night- and all-weather interceptor of the UARAF in the late 1950s and early 1960s. (David Nicolle Collection)

Following similar lines, EAF officers arrived in Damascus to take over the SyAAF. Makabri was left in his position and assigned the overall command of all the UARAF assets deployed in the Northern Province, as Syria was re-designated during this period. However, the mass of Syrian aircraft and personnel were transferred to units home-based in Egypt. This included all the radar-equipped MiG-17PFs and their pilots: they were assigned to the Inchas-based No. 31 Squadron, commanded by Squadron Leader Munir al-Garudy. In their place, the Egyptians deployed two squadrons of MiG-17Fs commanded by Squadron Leader Huseyn Sidqi. This is why the SyAAF operated radar-equipped MiG-17PFs for only a few months, and never managed to press its brand-new Il-28s into operational service before being integrated into the UARAF.[4]

NOCTURNAL RECONNAISSANCE

This is not to say that no Il-28s were ever operated from Syria: on the contrary. Before the UAR came into being, the UARAF and the DM1 had the idea of using the additional Il-28s to conduct a series of nocturnal reconnaissance missions over Israel in order to update their intelligence on the IDF/AF and its bases. At the time, the Air Force was in the process of working up Air Group 61 – a wing-sized formation consisting of about 50 crews and more than 40 aircraft organised into two squadrons equipped with Il-28 bombers and one with Il-28R reconnaissance aircraft. The biggest problem for Air Marshal Sidqi Mahmoud and his aides was that the DM1 and the UARAF had next to no knowledge about their possible opposition. Logically, they assumed that the interceptors of the IDF/AF would be the biggest threat for the Il-28Rs, and thus devised a plan in which the reconnaissance aircraft would launch from bases in Egypt late at night, penetrate the Israeli airspace at low altitude, take photographs and disappear into the darkness and at low level for a landing in Syria. Such tactics, the route and the speed of the Il-28Rs were expected to keep the aircraft and their crews safe.[5]

An Il-28 of the UARAF seen on landing at Beni Suweif AB. The suffix '1' to the code 'C' indicated that this was the second aircraft coded this way – most likely because the first was written off during the Suez War of 1956 or at a later date. As of the late 1950s, and through the 1960s, the type was operated by Nos. 8 and 9 Squadrons of Air Group 61. (From the film *Ismail Yassin in the Air Force*)

50 Il-28 crews completed their conversion and tactical training at Bilbeis AB in 1958, enabling the formation of Air Group 61: a wing-sized unit consisting of two squadrons equipped with Il-28 light bombers, and one with Il-28R reconnaissance bombers. (Nour Bardai Collection)

A group of Il-28 crews seen during the pre-mission briefing for one of the nocturnal reconnaissance sorties over Israel in 1958. (Nour Bardai Collection)

The first such operation is known to have been flown sometime in January 1958, apparently by two aircraft. One was underway ahead and acted as target marker: when in the vicinity of the target it would release several FOTAB-100 flare bombs to light the target for the second Il-28R, which would take photographs. Because the Israelis did not even react to the appearance of UARAF bombers, and as soon as the dust settled over the creation of the UARAF, the second mission was launched during the night of 2-3 April 1958. Once again, the IDF/AF was caught entirely unprepared and this time two pairs of Il-28Rs remained undetected until the target markers released their flare bombs. Although certainly encouraged, the UARAF returned to the Israeli skies only during the night of 24-25 November 1958, when two Il-28s that took off from al-Mezzeh AB in Damascus, are known to have flown down the entire length of Israel and were again detected only once the first of them dropped flare bombs.

14 TAMMUZ REVOLUTION IN IRAQ

In the eyes of the Arab public, during the Suez War of 1956 President Gamal Abdel Nasser won a major political victory over the former colonial powers, virtually ending their domination of the Middle East. A public image of Nasser was created that presented him as a person with all the attributes required for an Arab leader – qualifications which, supposedly, other Arab leaders lacked. The creation of the UAR bolstered such impressions by an order of magnitude, in turn prompting demands for various other predominantly Arab countries to join the Union. Lebanon was no exception. While created by the French as a Western stronghold in the Middle East, in which Christians were supposed to dominate the social, political and economic life, by the mid-1950s the country was predominantly populated by Muslims and they began demanding that Beirut follow Syria and join the UAR. When the government refused, mass demonstrations erupted, which gradually turned into riots, and riots into infighting between various Christian and Muslim militias. As the fighting spread from Beirut to Sidon and then Baalbek, the UAR Army deployed two brigades from the Northern Province into Lebanon, but then the Christian militias fought back to stop their deployment. By June 1958, fierce fighting was raging from Beirut to Tripoli.[6]

Alarmed, two other, British-established and supported governments in the Middle East – those of the Hashemite Kingdoms of Iraq and Jordan – decided to counter the emergence of the UAR by creating the Arab Federation. This was to have King Faysal II of Iraq as Head of State, one flag and one federal government with two regional administrations, and one army and air force. Correspondingly, in July 1958, several units of the Iraqi Army received orders to re-deploy to Jordan. This was the moment exploited by a group of Iraqi Army officers to launch a coup d'état in Baghdad: on 14 July 1958, Brigadier-General Abd al-Karim Qasim and his aide, Colonel Abd as-Salem Arif, led two brigades into the capital, secured the main radio station, announced a revolution and the downfall of the monarchy, and proclaimed a republic. Immediately after, the entire royal family, including King Faysal and Crown Prince Abdul Illah, the former Prime Minister of Jordan, and the entire government were arrested and subsequently executed.[7]

With Qasim promptly withdrawing Iraq from the Arab Federation and suspecting he might be removed in similar fashion by officers sympathising with Nasser, or some of the Iraqi units already deployed in Jordan, King Hussein then requested a British military intervention. In Beirut, President Camille Chamoun followed in fashion, requesting military assistance from the United States. Misinterpreting pan-Arabism for a joint conspiracy of the UAR and the USSR with the aim of imposing a Soviet-style 'communist' regime in Lebanon, and acting under the provisions of the UN Charta, both superpowers reacted promptly. From 15 July 1958, the USA deployed 14,000 troops to secure key points around Lebanon.

A long row of MiG-17Fs from the UARAF unit temporarily deployed at the former RAF Habbaniya AB in central Iraq, shortly after the 14 Tammuz Revolution. The aircraft in the foreground wears the serial number 2200. (Nour Bardai Collection)

During their stay in Iraq, Egyptian and Syrian pilots of the UARAF ran a joint exercise with No. 6 Squadron of the IrAF. This photograph shows a pair of Hunters (centre) with two MiG-17Fs. (David Nicolle Collection)

Similarly, at dawn of 17 July, the first of 2,200 British paratroopers deployed to Amman to secure the country for King Hussein, until several powerful Bedouin tribes pledged their loyalty.

FROM THE ROYAL TO THE IRAQI AIR FORCE

Although Lebanon and Jordan were thus prevented from joining the UAR, there was no way back for Iraq. Indeed, the local armed forces soon found themselves ordering Soviet arms. Established around a core of 600 former Ottoman officers of native origin whom the British permitted to return and settle down after the fall of the Arab Kingdom of Syria, the Iraqi armed forces have dominated the national politics in Baghdad right from the times the country was carved out on the basis of the Sykes-Picot Treaty. Following, what was effectively re-education, the officers were granted permission to establish an army for the purpose of helping the garrison of the British Army of India to suppress one armed uprising after the other – most of which were provoked by the fact that ever since the British enthroned King Faysal in Baghdad in 1921, his government was subjected to their tight control, ruled via a system of patronage favouring a small elite, and was repeatedly forced into signing treaties favourable to London. Correspondingly, the Royal Iraqi Air Force (RIrAF) came into being only once the British not only felt safely in control, but also needed a favour from Baghdad in 1931, and was then equipped with a miscellany of light aircraft, ill-suited to the requirements of military service. Nevertheless, and while not the first military flying service in the Arab world, the RIrAF did become the first official Arab air force staffed by natives. Over the following 25 years, it certainly became one of the most combat-experienced air forces of the world, too. Thanks to an extensive training program,

A MiG-17F (serial number 452) of the Iraqi Air Force in 1959. Notable is the fin flash in use until 1963, and the IFF antenna atop the fuselage. (Ahmad Sadik Collection)

The front section of the IrAF MiG-17PF 458, at the IrAF Museum in the Green Zone of Baghdad. The camouflage pattern was applied in Syria during the October 1973 War with Israel. (Photo by Mohammad Hassan)

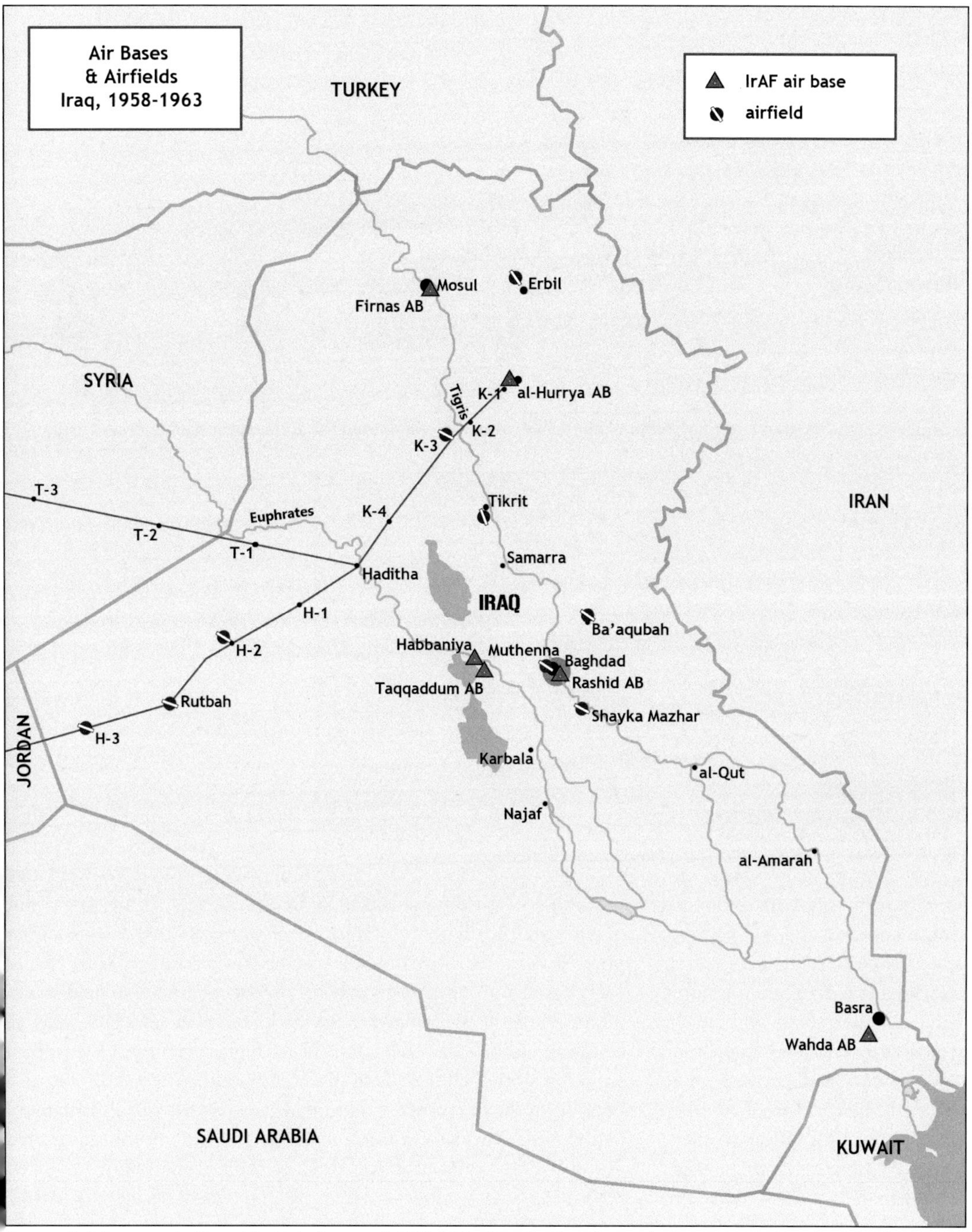

A map of air bases and other airfields in Iraq as of 1958-1963. (Map by Tom Cooper)

it grew rapidly during the 1930s, when it participated in operations against dozens of armed uprisings. While completely destroyed during the Anglo-Iraq War of 1941, it was slowly re-built during the 1940s, and then experienced another period of almost unlimited growth. By 1958, it was a service almost entirely equipped with British-made aircraft, and staffed not only by officers, but even NCOs all of whom underwent at least one training course in Great Britain, and some several such courses.

As soon as the new government established itself in power in Baghdad on 15 July 1958, the RIrAF not only lost the prefix 'Royal', but also received a new commander – Brigadier-General Jalal Jaffar al-Awqati – who was not only a pan-Arabist, but a staunch communist. Unsurprisingly, and in agreement with Qasim, the new leader in Baghdad, Awqati rushed to not only 'de-royalise' his force – which was to purge it of any officers considered loyal to the monarchy – but also invited the UAR to deploy elements of its armed forces to Iraq, to 'bolster the defences of the revolution'. This is how Cairo then came to order Squadron Leader Huseyn Sidqi to re-deploy one of his two MiG-17F units from the Northern Province to the former RAF Habbaniya, vacated by the British only recently. Next, Awqati hurried to establish good links with Moscow and is how it came to be that the first contract for the acquisition of Soviet-made combat aircraft by Iraq was signed only weeks later.

Although some Russian publications indicate deliveries of 19 MiG-15bis, what Iraq actually received was a miscellany of MiG-15UTIs, MiG-17Fs and MiG-17PFs, Il-28s, Antonov An-2s, Yakovlev Yak-18 basic trainers and Mil Mi-4 helicopters; the first of which were in the country by the end of the year. By March 1959, the IrAF re-equipped its No. 5 and 7 Squadrons with MiG-17s, re-established No. 8 Squadron (destroyed during the Anglo-Iraq War of 1941) equipped with Il-28s and introduced MiG-15UTIs as advanced trainers in its Air Force College.

KING-CHASERS

Meanwhile, in the Northern Province, Makabri, Sidqi and their UARAF units were constantly kept on their toes. With all the massive movement of US and British aircraft and troops around them, their first priority became the establishment of a functional early warning radar network. Consisting of Soviet-made P-8 radars and worked up with help of a group of about 30 Soviet advisors, this became functional during the summer of 1958, and was soon put to the test. Ever since the Suez War, the RAF – followed by the CIA – kept a watchful eye on developments in the Middle East with the help of high altitude reconnaissance overflights, undertaken by Canberras and Lockheed U-2s, respectively. The UARAF units now deployed at al-Mezzeh and Hama ABs launched numerous intercept attempts. However, by operating at altitudes of 15,240m (50,000ft) and higher, the intruders always remained outside the reach of MiG-17Fs. With President Nasser now widely perceived a 'protector of the Arabs' in public, British and American violations of the UAR airspace made UARAF commanders very nervous, and they began scrambling their interceptors whenever radars detected any kind of unannounced incursions. For example, on 14 November 1958, two MiG-17Fs intercepted a de Havilland Dove carrying King Hussein of Jordan for

Pilots of one of two UARAF MiG-17F units deployed in the Northern Province, scrambling for their aircraft in response to the appearance of another unknown aircraft inside their airspace, in 1958. (Photo by Osman Mahhmoud, via Nour Bardai)

A de Havilland Dove of the Royal Jordanian Air Force, seen around 1958. (John Fricker Collection, via Simon Watson)

a three-week vacation in Europe: the aircraft appeared unannounced inside UAR airspace because the commander of the RJAF forgot to forward its flight plan to Damascus. To make matters worse, its British pilot, Wing Commander Jock Dalgleish, RAF, then ignored signals from MiG pilots advising him to follow them: instead, he turned back south and began flying evasive manoeuvres. Uncertain about what to do, the two MiG pilots requested permission to open fire: the ground control remained silent. However, when they returned to the base, both were ostracised for failing to shoot down the aircraft carrying King Hussein. Dalgleish thus managed to fly the Dove undamaged back to Amman, where the news of this incident caused an eruption of 'anti-Syrian' protests.[8]

AIR WAR OF 1958

In the meantime, another crisis was brewing along the armistice lines in Sinai. Sporadic clashes there began erupting almost as soon as Israel withdrew from the peninsula in March 1957. Initially at least, they were primarily caused by relatives of about 1,300 Palestinians (including members of the Palestinian Border Guard and civilians), summarily executed by the IDF during the occupation, seeking revenge on the other side of the armistice line. However, with all the contingents of the United Nations Emergency Force (UNEF) deployed on the Sinai Peninsula and in positions along the armistice line, the situation on the ground became remarkably quiet, while all further clashes moved into the air.[9]

The deployment of the UNEF contingents was closely monitored by IDF/AF reconnaissance aircraft: subsequently, the Israelis continued violating Egyptian/UAR airspace on a regular basis, until the UARAF began to counter these with its interceptors. The logical result was a series of aerial clashes of growing intensity, during which each side operated differently. The Israelis would either send reconnaissance aircraft protected by interceptors into the enemy airspace, or their interceptors would try to lure UARAF interceptors into the Israeli airspace. Generally, the UARAF had standing orders not to violate the Israeli airspace, and thus its pilots were reacting to Israeli provocations: however, sometimes they would penetrate the Israeli airspace on their own initiative – whether in frustration caused by frequent Israeli violations, or out of enthusiasm and bravado. Radar coverage on both sides was rather poor, navigational instruments installed in aircraft of both parties were rather rudimentary by nature, and Egyptian, Syrian and Israeli pilots regularly made simple navigational mistakes. The mass of resulting clashes was characterised by high speeds and long-range shooting: with the majority of pilots still honing their skills, but also because of the previously mentioned peculiarities of the armament of the MiGs, next to nothing was hit, and thus nothing reported in public. However, pilot logbooks of the Egyptian pilots involved recorded a high number of active engagements between the UARAF and the IDF/AF. For example, the logbook of Abdel Moneim el-Shennawy – who flew MiG-15bis at the time – recorded his first air combat with Mystère IVAs on 1 September 1957 (in the jet with serial number 2651), and then additional engagements on 4 and 5, 8, 9, and 10, and 19 April, and on 22 and 26 May, 16 June, 30 July, 24 August, 28 October, and 17 November 1958. As the intensity of aerial engagements continued to grow, Air Marshal Sidqi Mahmoud decided to bolster the UARAF deployment in Sinai. By November 1958, el-Arish AB was expanded to enable a permanent deployment of a full squadron of MiG-15bis or MiG-17Fs. Furthermore, disused British-constructed airfields at Bir Thamada, Bir Gifgafa and el-Sur all received hardened and extended runways to enable operations of jet fighters.[10]

Accustomed to their relatively luxurious life at bases close to Egypt and Syria's major urban centres, personnel of UARAF units now temporary deployed at air bases 'in the middle of the desert' found itself working under harsh and primitive conditions, but – based on experiences from 1956 – everybody was in high spirits. The Israelis did not let them wait for long. On 11 December 1958, a pair of brand-new Dassault Super Mystère B.2 (SMB.2) interceptors

A MiG-15 of the UARAF diving during an aerobatic exercise. Although it had proven inferior in speed in comparison to the Mystère IVA during the Suez War of 1956, the type still possessed a slight advantage in the top operational altitude. With this being considered of paramount importance well into the early 1960s, the Egyptians continued using it as a frontline interceptor. (Nour Bardai Collection)

Abdel Moneim el-Shennawy with a MiG-15bis at Kabrit AB, in 1958. Notable is the unit insignia – that of the MiG Operational Conversion Unit (see colour section for details) – on the nose of his jet, and the slipper-type drop tank under the wing. (Tarek el-Shennawy Collection)

attempted to ambush a pair of MiG-15bis, but the Egyptians managed to avoid them. Because of this provocation, and because of reports about a build-up of Israeli ground units along the armistice line, the DM1 warned about an imminent Israeli invasion. Taking this threat very seriously, Field Marshal Amer ordered reinforcements for the Army units in Sinai, while Air Marshal Sidqi Mahmoud once again sent his Il-28Rs to investigate. The first nocturnal reconnaissance mission over Israel undertaken by the crews of Air Group 61 during this period is known to have taken place on 14 December, when two Il-28Rs made a shallow penetration in the south, to take photographs of IDF units deployed along the armistice line. Two days later, the UARAF then deployed five Il-28s for a much more ambitious undertaking: while one was to act as a decoy, the other four were to fly the full length of Israel from south to north before landing in the Northern Province, to photograph Ramat David, Tel Nov and Hatzor ABs. By sheer accident, it was only during this mission, flown on 16 December 1958, that the IDF/AF – finally – managed to launch an interception in time. A lonesome Meteor NF.Mk 13 eventually found one of the Il-28Rs and disturbed its mission, painting it with its radar and thus forcing the pilot to take evasive action. However, immediately after, the Israeli fighter was successfully distracted by the decoy Il-28: this climbed to a higher altitude and allowed itself to be detected by the Israeli early warning radars, prompting the IDF/AF's ground control to re-direct the Meteor in that direction. Eventually, all the UARAF aircraft distanced without further problems.[11]

The results of the nocturnal reconnaissance operations by the UARAF over Israel at this point in time remain unknown: whatever photographs the Il-28Rs might have taken, it remains unclear if these were even forwarded to the DM1, or if the later made any use of them. The only detail known is that during the mission on 16 December 1958, one of the Il-28s passed directly over Tel Nov AB, but its crew was late in activating its cameras and thus returned empty handed.

Certainly enough, after this and with the UAR Army units moving into Sinai, the Israelis began to investigate on their own and on 20 December 1958 their reconnaissance aircraft provoked a clash between two UARAF MiG-15bis from el-Arish and four SMB.2s deep over Sinai. The MiGs were trailing the lead Israeli pair, which acted as bait, when the rear pair attacked them from above, claiming one UARAF jet as shot down near Bir Lahfan, fully 65 kilometres inside Sinai. No MiGs are known to have been lost on that day, nor during this entire period. Perhaps because of this, sporadic border clashes went on for days longer, including air combats on 23 and 29 December 1958. Indeed, the tensions lessened and the two parties disengaged only after two further air battles. The first took place on 8 January 1959, when a 'finger four' formation of MiG-15bis led by Ahmad el-Dirayni, including Medhat Enan as Number 2, Abdel Moneim el-Shennawy as Number 3, and Gamal Afifi as Number 4, clashed with four Mystère IVAs. The Israelis attacked attempting to exploit the superior speed and climb capability of their jets, but Dirayni forced one to overshoot. When the Israeli attempted to evade by a hard break to the left, his jet was hit by several 23mm shells, spiralled out of control and hit the ground. Meanwhile, Shennawy outmanoeuvred the other Israeli who attempted to disengage in a high-speed climbing turn, the Egyptian fired a well-aimed volley

Gamal Afifi in the cockpit of the MiG-15bis 2651, flown by Abdel Moneim el-Shennawy on 11 and 23 December 1958. (Abdallah Emran Collection)

Ahmad el-Dirayni in front of the 'Black Leopard' insignia applied on the MiG-15bis 2651 flown by el-Shennawy during several air battles with Israelis, in November-December 1958 (see colour section for further details). Note his – then brand-new – 'helmet', including hard-encased earphones, issued to UARAF pilots in 1958. (Nour Bardai Collection)

Captain Omar Farouk Abdel Salam in the uniform of the UARAF. (Nour Bardai Collection)

in front of him, causing the Mystère IVA to explode. Both Israeli losses were recorded by the gun-cameras carried by all MiGs and confirmed by UN observers. The second air battle followed less than a month later, in early February 1959, when Omar Faouk Abdel Salim claimed another Israeli fighter as shot down and brought

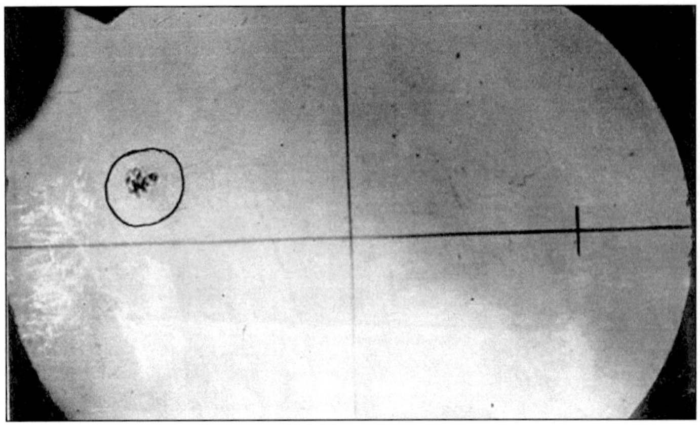

Two stills from Abdel Moneim el-Shennawy's gun camera film, showing him aiming in front of the Mystère IVA as this was climbing to the right, and then the explosion of the Israeli jet caused by his shells. (Nour Bardai Collection)

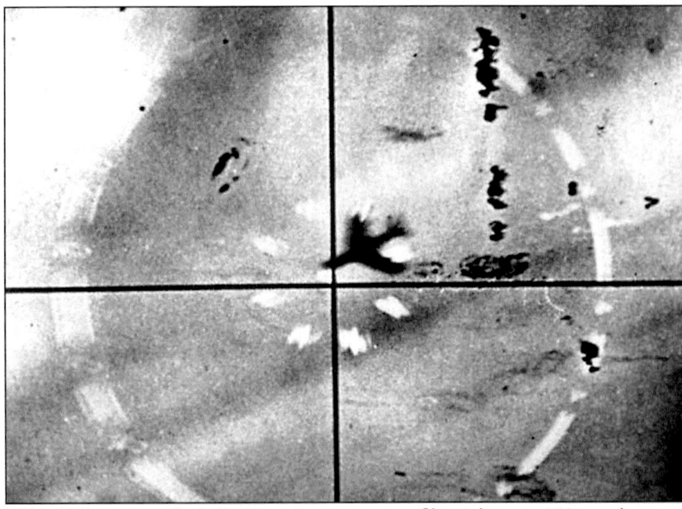

This still from Farouk el-Ghazzawy's gun camera film, taken on 4 November 1959, leaves no doubt about the reason for the ejection of Israeli Captain David Ivry (later appointed a commander of the IDF/AF). (Lon Nordeen Collection)

back a gun camera film confirming his claim. Available Israeli accounts not only never mentioned the majority of the above-listed engagements, but no losses either. In another incident, a short clash between two UARAF MiG-17Fs and two SMB.2s on 4 November 1959, they did admit one loss, but stated that their pilot had ejected because his aircraft entered a spin during a hard break – although the UARAF released a clear gun camera sequence showing the jet in question under fire from the MiG piloted by Farouk el-Ghazzawy.

In the light of these experiences – widely published in the UAR, but entirely ignored elsewhere – the senior leadership of the UARAF concluded that the force was superior to the IDF/AF. Unsurprisingly, over the following years, Cairo became convinced it would dominate the skies in any bigger conflict.

WATER WAR, ROUND 1

At the end of the Palestine War, several UN-controlled DMZs were created along the ceasefire lines between Israel, Jordan and Syria: officially at least, the ground there belonged to nobody, and none of the parties was to deploy any kind of military units there. Starting in 1951, Israel began violating the DMZs vis-à-vis Syria by deploying military units within, supposedly necessary to protect construction companies involved in the draining of Lake Hula (part of the Jordan River system). Whenever opportune, the IDF units in question fired at nearby Syrian villagers. Knowing it was too weak to confront Israel frontally, Damascus ordered a deployment of its Army units *around* the DMZs but refrained from entering them or shooting back at the Israelis.[12]

Over the following years, both sides developed their own designs for the development of the Jordan Plain: in 1950, Israel began planning its National Water Carrier (NWC), intended to divert water from a point on the Jordan River near the B'not Yaakov Bridge – inside one of the DMZs with Syria, and thus in violation of the armistice agreements from 1949 – while two years later, Jordan and the United Nations Relief and Works Agency (UNRWA, a body originally created to provide relief and human development to the Palestinian refugees) agreed to launch the construction of a canal that would siphon water out of the Yarmouk River northeast of its confluence with the Jordan, and transport it south, parallel with and just to the east of the River Jordan, where hundreds of thousands of Palestinian refugees were concentrated. Tensions over both designs were high before any construction work began, and thus the Eisenhower administration urged the UN to take a lead and design a unified plan acceptable for an equitable apportionment and utilisation of the River Jordan and its tributaries among the riparian countries. Strongly opposing any such ideas, Israel took care to secure US funding for the construction of its NWC, and on 2 September 1953 launched the related work inside the DMZ with Syria. Protests from UN representatives were ignored: indeed, the Israelis only accelerated the work. Still in no position to fight a war over water, Damascus could find no other solution than to bring the matter to the UN Security Council – eventually prompting Washington to react: when there was no reaction from Israel, on 18 September 1953, the Eisenhower administration stopped providing all financial aid to the country. About a month later, the Israeli government agreed to stop the diversion project: however, as soon as the US aid resumed, the Israelis completed their unilateral water project – this time with the help of German reparations for the Holocaust.[13]

Following Israel's refusal to cooperate, Arab countries were not the least keen to become involved in any kind of projects certain to make a major contribution to Israel's economic growth. In 1955, the Arab League decided not to ratify the joint plan eventually developed by the USA, even if adhering to its technical details. Instead, Syria and Jordan sought to divert the Jordan River headwaters to the Yarmouk River. This complex scheme proved not only technically difficult, but expensive and next to impossible to realise. Nevertheless, construction work commenced two years after Egypt and Syria formed a union.

TEWFIK WAR

In early 1960, the construction work on the Yarmouk diversion project was initiated near the village of Tewfik, three kilometres inside the Northern Province (and thus well away from the armistice line). Complaining that the Syrians were attacking their settlements and the NWC, the IDF reacted by deploying troops within the DMZ and shooting at anything that moved on the other side. When the UAR Army did not respond, the IDF raided the construction site, on 31 January 1960, destroying equipment and housing under the pretext of a 'retaliation raid'. The following morning, fighters from both sides were airborne and there were several engagements,

From left to right: Captain Neqola Qori, 1st Lieutenants Mowafaq Daghestani, Samir Mohammad al-Masry, and Adnan Haydar after their successful engagement with the Israelis on 31 January 1960, when Daghestani and al-Masry each claimed one SMB.2 as shot down. (Nour Bardai Collection)

1st Lieutenant Muwaffaq Daghestani and one of his wingmen reconstructing the air combat from 31 January 1960 to the representatives of the international press. (Nour Bardai Collection)

but all ended inconclusively. This changed around the noon, when a formation of four UARAF MiG-17Fs, led by Captain Neqola Qori, with 1st Lieutenants Adnan Haydar, Muwaffaq Daghestani and Mohammad al-Masry, clashed with a section of SMB.2s. What exactly happened during the following dogfight remains unclear. According to Israeli publications, there were only two Super Mystères involved, and one MiG was damaged without any losses; however, according to Egyptian and Syrian sources, not only did all four MiGs return to their base safely, but Daghestani and al-Masry each claimed a kill against an SMB.2 and brought back gun camera films confirming this.

MiG-17Fs piloted by el-Kossary and Massekh during their engagement with SMB.2s, on 19 October 1960. Sadly, their serial numbers were removed by the military censor, but the prominent IFF antenna on the back of the aircraft, and full set of UARAF markings are all clear to be seen. (Nour Bardai Collection)

A MiG-15bis (serial number 2703) of the Fighter Conversion Unit in the process of tucking in its undercarriage after take-off. The unit regularly rotated small detachments of pilots and aircraft to el-Arish AB, and these saw much action through 1958 to 1960. (Photo by Tahsin Zaki, via David Nicolle)

After this clash, the IDF began massing ground troops along the armistice line – prompting the DM1 to sound another alarm in Cairo. On 18 February 1960, Nasser reacted by ordering a 'silent' mobilisation of the UAR Army: one not publicly announced. For four days, Egyptian troops streamed into Sinai without the AMAN having even a trace of a clue about what was going on. During the following days, UARAF Il-28Rs flew at least two nocturnal reconnaissance sorties over northern Israel: one along the route Safad – al-Jauna – al Khalisa, and another along the Jordan River with a 'wide circle' over Kfar Shamir. Both caught the IDF/AF entirely unprepared and not one Israeli interceptor was scrambled.[14] It was only once the Israelis reacted to this overflight by sending a pair of Sud-Ouest Aviation (SNCASO) S.O.4050 Vautour II reconnaissance fighters over Sinai that they detected the massive presence of enemy troops close to the armistice line. In a state of deep shock, the IDF ceased raiding Tewfik and instead scrambled to mobilise in the south. By then, it was already too late: not intending to invade, but only to demonstrate power and lessen the Israeli pressure in the north, Nasser – who *was* a former military officer, but now primarily considered himself 'a politician playing a game of chess' – ordered his army back to the Canal Zone. Unknown to the Egyptian leader, and while there was little doubt that his message was delivered, this affair was to play an important role seven years later.

In another little-known encounter between UARAF and IDF/AF jets over the Sinai Peninsula, on 19 October 1960, Captain Hassan el-Kossary (left) and 2nd Lieutenant Ali Massekh, attempted to intercept an Israeli Vautour. While the escorting SMB.2s claimed to have shot down one of the two MiGs, el-Kossary and Massekh each claimed one SMB.2. Apparently, all the claims in question were exaggerated. (Nour Bardai Collection)

OUTPOST IN THE DESERT

Through March 1960 the UARAF lessened the intensity of its operations along the armistice line with Israel and limited these to the regular combat air patrols by MiG-17 units – mostly drawn from Air Group 2, which at the time included Nos. 2, 5 and 18 Squadrons – rotated to el-Arish every four to six weeks. On the contrary, still sensitive to the Egyptian mobilisation it missed the month before, the IDF/AF continued flying regular reconnaissance sorties over the Sinai Peninsula for months longer. All such missions were rigorously protected by interceptors, and thus occasional air battles remained

A MiG-17F of No. 18 Squadron, UARAF, seen over the mountains of central Sinai. Initially deployed in an on-off basis, this unit was eventually home-based at el-Arish AB. Depending on the flight in question, rudders and wingtips of its aircraft were decorated with red or blue checkerboard patterns. (Photo by Abdel Moneim al-Tawil)

Taher Zaki with MiG-17F serial number 2038 at el-Arish AB during one of the deployments in the early 1960s. (Taher Zaki Collection)

unavoidable. During his next tour at the lonesome Egyptian 'outpost in the desert', Abdel Moneim el-Shennawy and his wingman clashed with a pair of SMB.2s on 30 May 1960, albeit without success. On 19 October 1960, Israeli SMB.2s ambushed a pair of MiG-17Fs trailing a Vautour and claimed one as damaged. On 28 April 1961, two MiGs managed to force an Israeli aircraft to abort its mission, but in turn were attacked by SMB.2s while approaching the enemy airspace. In the ensuing dogfight one of the Egyptians and one of the Israelis fell into a spin. While the Israeli apparently managed to recover his jet in time, the Egyptian pilot was forced to eject and his aircraft crashed Nizzana. The Egyptian pilot managed to dash to the other side of the armistice line and thus evade capture, but the final moments of this clash were witnessed by UN troops of the Yugoslav and Canadian contingents, and thus reported.[15]

TENSIONS WITH JORDAN[16]

King Hussein of Jordan felt under severe pressure from the UAR not only through 1958, but for a while longer, even though Nasser eventually issued guarantees for the sovereignty of his country. On 29 August 1960, a bomb exploded in the office of his Prime Minister, Haza'a Barakat al-Majali, killing him and 11 other people. Suspecting 'Syrians' were responsible for this act, Hussein decided to launch a reprisal attack – in cooperation with Israel if necessary. Therefore, the three fighter-bomber units of the RJAF existent at the time – two equipped with Hunters and one with Vampires – were redeployed to the H-5 airfield in eastern Jordan and briefed for attacks on UARAF air bases in the Northern Province. Very little detail about the plan for that operation is known, except that Vampires of No. 3 Squadron were to strike al-Mezzeh and Aleppo, while No. 1 Squadron was to hit Dmeyr AB. For days after, Jordanian pilots flew practice sorties over similar distances, and trained with use

Piloted by Adnan Malawi, this MiG-17F made a belly landing southwest of Aman on 4 October 1960. Judging by its serial number – 48 – it was one of 60 MiG-17F/PFs originally purchased by Syria directly from the USSR in October 1956 and delivered in 1957. By 1960, it received the full UARAF insignia, including classic national markings with two green stars and identification stripes around the rear fuselage and wingtips. (Greenhill via Patricia Salti)

of live ammunition, including 30mm rounds for their internal cannons and unguided rockets. The planning and preparations were largely completed by 8 September 1960, by when two brigades of the Jordanian Army were concentrated on the border, together with all of the armour and artillery support that was available. Eventually this operation was cancelled, under British pressure, several days later.

The UARAF seems to have got the wind of such preparations, and MiG-17Fs from one of the two units commanded by Sidqi flew intensive patrols along the border to Jordan. These resulted in two non-combat-related losses. On 27 September 1960, one MiG came down in the desert of northern Jordan: the pilot was injured but subsequently returned to the UAR. Only a few days later, on 4 October 1960, the second MiG-17F – which, according to the Jordanians, was equipped for photo-reconnaissance – made a belly landing southwest of Amman. Its pilot – variously named as Adnan Madani or Adnan Malki, depending on source – first said he did not intend to defect, but then requested political asylum. While waiting for his fate in the custody of the RJAF, he eventually committed suicide. His jet was recovered and thoroughly examined by 'American and British representatives', but its ultimate fate remains unknown.

Two Soviet representatives on arrival in Egypt, together with two Egyptian officers, and a Soviet Air Force Il-14 transport to their rear. Due to a de facto 'cold war' between Cairo and Moscow of the time, relations between the UAR and the USSR were almost exclusively formal. (Albert Grandolini Collection)

Photographs of original Egyptian MiG-19Ss from the early 1960s remain extremely rare. This example from No. 27 Squadron was photographed while passing low over Fayid AB. (David Nicolle Collection)

THE MIG-19 FAILURE

During the clashes with Israel in 1958-1960, the UARAF came to the conclusion that not only were its MiG-15bis outflown by the Mystère IVAs, but also that MiG-17Fs and MiG-17PF were underdogs compared to the French-made SMB.2s. Therefore, and although burdened by the debts from earlier acquisitions, Cairo began demanding deliveries of more advanced fighters from the USSR. For not directly related reasons, negotiations were anything but easy: immediately after the 14 Tammuz Revolution in Iraq, Nasser hastened to Moscow to seek Soviet military action in the Middle East, with the aim of protecting the 'revolution' in Iraq from possible Western interference. He repeated similar requests in the light of the US and British interventions in Lebanon and Jordan, respectively. However, fearing their unlimited support might prompt the 'impulsive' Egyptian leader into launching unilateral military actions that might cause a confrontation with the USA, Moscow refused to move.[17] The Soviet government soon saw its decision confirmed: when the new Iraqi leader began suppressing the influence of the Iraqi Communists, Nasser felt challenged to a degree where he renewed the persecution of Leftists in the United Arab Republic too. This was something the leadership of the Soviet Union was not ready to tolerate: because it had at least one firm ally in Baghdad, in the form of the IrAF commander Awqati, it reacted with public critique of Cairo, eventually provoking a 'cold war' between the USSR and the UAR. Certainly enough, reason prevailed: having no other arms supplier, Nasser continued purchasing Soviet arms, and – in hope of retaining at least some influence in Cairo – the Soviets continued delivering. Indeed, the two countries introduced a pattern in which a new large arms deal was signed every second year. The first of these was penned in 1959, and envisaged deliveries of 56 MiG-19S supersonic interceptors, and Soviet assistance in constructing five air bases in the Sinai Peninsula.[18]

From right to left: Major Alaa el-Din Barakat (CO No. 27 Squadron and future commander of the EAF), Captain Galal Abdel-Alim, Captain Farouk Abdel Latif, and Captain Abd el-Moneim el-Tawil in front of the fin a of MiG-19S of the UARAF in 1962. (David Nicolle Collection)

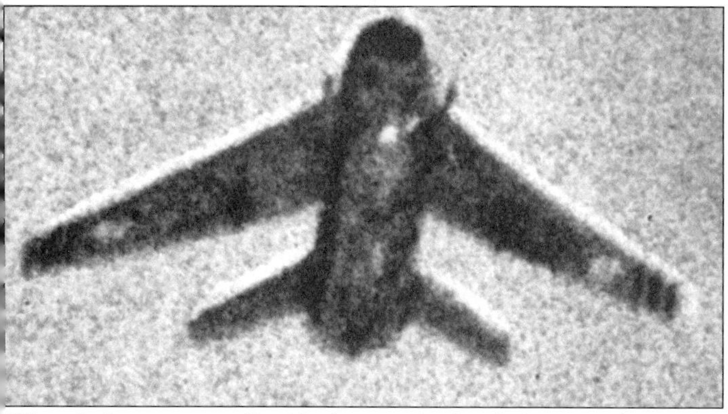

Although of very poor quality, this photograph taken during a parade in Cairo in 1962 is of particular significance because it is the first to clearly show identification stripes applied around the wingtips of this MiG-19S – in turn showing that the tradition of applying these was continued on jets of Nos. 27 and 29 Squadrons. (Abdallah Emran Collection)

Nasser announced the acquisition of new fighters only in May 1960, about a month before the first group of UARAF pilots travelled to the USSR for conversion training. On their return, they created the core of two newly established units: the Fayid-based No. 27 Squadron, commanded by Ahmad el-Dirayni and the Hurghada-based No. 29 Squadron, commanded by Alaa el-Din Barakat, both declared operational in 1961. However, during initial operations, MiG-19s proved to be nothing but trouble. Production of this type ended in 1959 and in the USSR it was replaced by superior designs. Correspondingly, what the UARAF received were second-hand aircraft from early production, taken out of storage at the Novosibirsk-Yeltsovka Works. Problems with them soon became obvious during the conversion training of Egyptian pilots in the Soviet Union, when technical unreliability of the type resulted in the loss of Egypt's most successful fighter-pilot of the Suez War: on 14 November 1960, Nazih Khalifa was killed during a nocturnal test flight in MiG-19S.[19]

Arriving without the necessary overhauls and operated in the hot climate of the Middle East, second-hand MiG-19s began suffering from numerous problems, most critical of which were – frequently worn out and certainly flimsy – hot air pipes that ran close to the hydraulic tanks, and the two engines tending to overheat each other. Finally, the overall manufacturing quality of the MiG-19 was so low that the airframe tended to bend in high-g turns. Unused to operating such a problematic type, the UARAF experienced such attrition that all the plans for another order were abandoned, while the rapid attrition to accidents of all kinds reduced the two operational units to just one in a matter of two years: by 1963, the surviving MiG-19s and personnel of Nos. 27 and 29 Squadrons were merged into No. 20 Squadron.

IRAQI EXPERIENCES[20]

Meanwhile in Iraq, the euphoria over the UAR and the 14 Tammuz Revolution had hardly settled down when the new government began persecuting any person or movements either in disagreement with it or proving disloyal. Before long, purges of the armed forces reached such proportions that a number of IrAF units were down to a mere handful of qualified pilots. Eventually, to keep officers and other ranks under control without having to distance them from the service, Awqati was forced to re-organise the entire service according to the political orientations and preferences of its members. Correspondingly, officers and other ranks either being pan-Arab nationalists, or even royalists, but not involved in any kind of plots against the government, were concentrated within No. 1 Squadron (which continued flying Hawker Furies), and No. 6 squadron (equipped with Hunters); Communists were concentrated within Nos. 5 and 7 Squadrons; and all others were re-distributed to Nos. 3 (transport) and 4 (helicopter) Squadrons. Furthermore, and in attempt to lessen Moscow's worries about Qassim's persecution of Leftists, Awqati took care to order 50 MiG-19 supersonic interceptors with a declared intention to establish a sort of 'flying praetorian guard', consisting of three units that would be responsible for the defence of the government.

The future Iraqi MiG-19 pilots underwent their conversion courses in the USSR, while the Soviets took care to rapidly deliver the jets in three variants, including:

- 30 MiG-19S, equipped with three 30mm Nuddelmann-Rikter NR-30 guns and two ORO-57K pods for 57mm S-5K unguided rockets
- 10 MiG-19Ps, equipped with the RP-1 Izumrud radar and only two NR-30 guns
- 10 MiG-19PMs, equipped with the RP-1 Izumrud radar and four Kaliningrad K-5M (ASCC/NATO-codename 'AA-1 Alkali') beam-riding, guided air-to-air missiles instead of cannons.

The majority of these aircraft reached Basra by ship in late 1959 – and thus long before any were delivered to Egypt – and were then trucked to Wahda AB, for assembly and test-flying by Soviet advisors. However, before long, the Iraqis realised that not

The only clear photograph of operational MiG-19Ss of No. 9 Squadron, IrAF, from 1959-1963 shows the examples with serial numbers (from left to right) 499, 498 and 502 while overflying Baghdad. Notable is their contemporary fin flash and application of the national insignia on the rear fuselages and near wingtips. (Nour Bardai Collection)

only were the aircraft in rather poor condition. Moreover, by that time the IrAF was lacking the personnel necessary to establish the three units planned to be equipped with the type. Eventually, the Iraqis took up only 16 MiG-19Ss, and pressed these into service with the newly established No. 9 Squadron. All the other MiG-19s were neither accepted nor paid for by Baghdad and remained stored inside a hangar at Wahda AB for several years longer. The IrAF thus did not become the first air force in the Middle East to introduce guided air-to-air missiles to service. Nevertheless, No. 9 Squadron was officially declared operational on 11 June 1961.

It was under similar circumstances and because of the worsening financial situation of Iraq – further complicated by Qasim nearly provoking a showdown with Great Britain over Kuwait in the late summer of 1961 – that Awqati was prevented from building-up three bomber-squadrons for the Air Force. After working up No. 8 Squadron with Il-28s, Awqati was forced to reduce the acquisition of Tupolev Tu-16 bombers – in 1960-1961 – to only 10 aircraft. The unit equipped with them, the newly established No. 10 Squadron, became operational in 1962.

5

NEW GOVERNMENTS, OLD HABITS

The creation of the UAR and the 14 Tammuz Revolution in Iraq of 1958 both created immense expectations in the Arab public. Ironically, while so seriously feared and opposed by Western powers and Israel, neither proved successful – largely due to the leaders involved. Within just three years they provoked the first of a chain of counter-coups which – ultimately – shattered all dreams of Arab unity. With much of the Arab world thus thrown into a state of chaos, the acquisition of further combat aircraft of Soviet origin nearly stopped – before being accelerated to unprecedented dimensions.

LAST UARAF RECONNAISSANCE OPERATION OVER ISRAEL

Late in the evening of 20 September 1961, the UARAF launched its most significant nocturnal reconnaissance sortie over Israel. Once again, the objective was to photograph the three major bases of the IDF/AF. The new element was that an additional aircraft was to overfly the Israeli nuclear complex in Dimona. Once again, four Il-28Rs were involved: they penetrated the Israeli airspace coming from the south, while a single Il-28 acting as a decoy did so from the north. As in all previous cases, the mission caught the Israelis entirely unprepared and then the IDF/AF failed to scramble in time and conduct an effective intercept. Eventually, only a single Vautour interceptor became airborne: this caught one of the Il-28Rs after it had completed its reconnaissance run and climbed over one of the hills in northern Israel but was then successfully distracted away by the timely reaction of the decoy Il-28. Wasting precious time to re-position for attack, the Israeli crew then approached the UARAF bomber to fire a quick burst from its 30mm guns but was in turn scared away by a well-aimed stream of tracers from the rear barbette as the Egyptian tail-gunner fired back. However, despite sound planning and execution, Egyptian sources indicate that the results of this operation were of very poor quality. Unknown to everybody involved, this was to become the last such operation ever undertaken.[1]

COUNTER-COUP IN DAMASCUS

If the creation of the UAR was met with much enthusiasm all over the Middle East in 1958, during the following three years the Egyptian tendency to patronise the Syrians created it plenty of enemies, too. This was not only the case in the Northern Province, but within the UARAF as well – where some units saw almost constant involvement in confrontations with Israel, while others saw next to no action. The latter seems to have been true at least for the Syrians assigned to the MiG-17PF-equipped No. 31 Squadron. While this unit is known to have lost one of its MiG-15UTIs during an air combat exercise in May 1960 (Abdel Moneim at-Tawil ejected safely), Hafez al-Assad – who was assigned as its commander around the same time – still found enough time to busy himself with clandestine political activity. Indeed, he joined several officers of

Two Il-28 crews preparing for their nocturnal excursion into Israeli airspace. (Nour Bardai Collection)

Ground crews inspecting an Il-28 on return from a nocturnal reconnaissance sortie over Israel in September 1961. Sadly, the censor has removed the aircraft's code (probably 'G1') and the unit number from the lower part of the squadron crest, which is otherwise clearly visible. This was probably the decoy Il-28 from that mission. (Ahmed Keraidy Collection)

Haider el-Kasbary, the MiG-17 pilot that flew the strike on the residence of Field Marshal Amer early on the morning of 28 September 1961, during the counter-UAR coup in Damascus. (Nour Bardai Collection)

Alawi origin – including Salah Jadid (who graduated from the Military Academy in Homs together with Hafez al-Assad, in 1950), Muhammad Umran (an Army officer specialised in armour), Abd al-Karim al-Jundi (technical officer of the SyAAF), and Haider el-Kasbary (MiG-17 pilot) – to plot not only a termination of the UAR, but also the dominance of the traditional civilian establishment in Damascus. To make matters worse, through 1961 President Nasser began playing directly into their hands: he tightened the state control over the Syrian economy, seized grain stocks, abolished regional government while appointing his Minister of War, Field Marshal Abdel Hakim Amer the 'Inspector General' in Damascus, and began planning a purge of the remaining Syrian officers. What were originally only rumours about the latter intention, became clear when Major-General Wadiyah Makabri was dismissed and replaced by Brigadier-General Muwaffaq al-Assasa, renowned as Nasserist. Unsurprisingly considering all these actions, by August 1961 a major political struggle ensued, resulting in demonstrations in several Syrian cities.[2]

In the early hours of 28 September 1961, Lieutenant-Colonel Abd al-Karim an-Nahlawi, the chief of Amer's bureau in Syria, secured al-Mezzeh AB, before leading the air defence troops of the local UARAF garrison – reinforced by an armoured battalion of the Army – into downtown Damascus, to seize the radio station and establish a number of checkpoints. A sole MiG-17F piloted by Kasbary was then scrambled to rocket Amer's residence: when Squadron Leader Sidqi appeared at the base, he was shot and badly injured by Syrians that suspected him of trying to steal one of the jets and deploy it against them. Field Marshal Amer was arrested slightly later, together with numerous Egyptian and Syrian officers loyal to Cairo – including the commanders of an Egyptian-staffed artillery brigade that attempted to intervene. Refusing to negotiate, Nasser ordered the UARAF to deploy a company of 120 paratroopers in combat jumps outside Aleppo and Latakia: by then, most of the UAR Army units in the Northern Province had sided with coup plotters and all the paras were arrested only minutes after landing, forcing the President to call off the operation. At 17.20hrs local time, Amer and a group of top Egyptian officers and loyalists were put on an airliner to Cairo. Following further negotiations, 870 Egyptian officers were sent home on 2 November 1961, in exchange for 960 Syrians that served in Egypt.[3]

Fearing repercussions because of Nasser's popularity, the coup plotters initially tried to maintain the UAR. However, when negotiations with Cairo ended nowhere, a break-up of the union was unavoidable. Syria was reinstated as an Arab Republic and even re-introduced its old flag in green, white and black, while Egypt retained the official designation UAR: correspondingly, the Egyptian Air Force also retained the title UARAF for a decade longer.

REBUILDING THE SYAAF

With the Egyptians out of the country, in September 1961 the Syrians scrambled to re-establish the SyAAF. To say this task was problematic would be a massive understatement: during the three years of the union, the country had received its first functional early warning radar network. However, in turn nearly everything useful was removed from local air bases, including not only the mass of aircraft and equipment, but even maps and furniture. Indeed, even the search for a new commander of the service took some time, until a solution was found by re-appointing Major-General Makabri on 19 October 1961. The 'new old' commander first took care to re-launch the training of Syrian pilots and ground personnel. In this regard, the Syrians had to start from the scratch: to their luck, during the times of the UAR, the UARAF had concentrated all of the remaining Egyptian and Syrian Chipmunks in Syria: these were gradually overhauled, together with several Spitfires and Meteors left from the 1950s and deployed to serve as training aircraft at the re-established Air Force College, now temporarily based at Minakh AB, north-west of Aleppo. With this proving insufficient to re-launch conversion training of new pilots to jet fighters, by the end of the same year a group of about 20 young fliers was sent for a conversion course to MiG-17Fs at Lugovaya in the USSR.[4]

For the time being, the combat element consisted of about 40 MiG-17Fs, a few MiG-15UTIs, and four Il-28s: the latter were still at al-Mezzeh as of 28 September 1961 by sheer accident, following their nocturnal reconnaissance sortie over Israel. However, the units operating them were critically short of almost everything – including pilots and ground personnel. The principal reason was that in the months after the coup the re-established armed forces became fractured along political, regional and religious lines. Indeed, Damascus was wracked by a series of coups and counter-coups, which culminated in four attempts initiated between 28 March and 2 April 1962. If there was anything still unifying about the Syrian armed forces of this period, then it was the wish to retain control over the government and prevent a return to a functional democracy. Therefore, on 1 April 1962, the top brass of the armed forces met in Homs, and agreed to exile not only the plotters of 28 September, but all the pro-Nasser officers and everybody who cooperated with the Egyptians during the time of the UAR, including the Ba'athists. Finally, the officers agreed to reinstall a civilian government led by Nazim al-Qudsi – once an ally of al-Atassi – as President. Amongst the SyAAF pilots that had to flee the country as a consequence was Hafez al-Assad: he escaped over the border to Lebanon, only to find himself arrested, repatriated to Syria and jailed.

MIRACULOUS RECOVERY

Whatever was going on in Syria in 1961-1962, there was one positive result: raising the country almost like a phoenix out of the ashes, the administration of President Qudsi worked hard on wide ranging economic and political reforms: banks, insurance and industry nationalised during the times of union were returned to their owners, and good relations were established not only with the UAR, but also Jordan, Lebanon and Saudi Arabia. Indeed, for a while at least, Qudsi had managed to restore contacts to Great Britain and the USA, prompting President John F Kennedy to label him a 'friend' of the United States. Encouraged by such developments, Makabri initially developed hopes to re-equip the SyAAF with Western-made aircraft: through the autumn of 1961 his emissaries toured Europe, investigating possible options. However, the Western European powers showed no interest. Thus, the delegation ended up in Prague to – following appropriate Soviet permission – sign a contract for delivery of spare parts for the Soviet-made equipment already in service. In February 1962, a delegation from Damascus then visited Moscow to re-establish relations and initiate negotiations for a new arms deal. Talks went on for several months, until on 19 June Damascus and Moscow announced the signature of a contract for – amongst others – 34 MiG-21F-13s and four two-seat MiG-21U conversion trainers. The deal was very favourable for Syria, with first payments due only in 1965, and further instalments spread over 10 years, at very low interest rates, and payable with Syrian produce valued in British pounds. Moreover, the Soviets agreed to provide

One of only four Il-28R reconnaissance bombers left behind by the UARAF in Syria. While Egyptian sources assessed the re-established SyAAF as not in a condition to make such aircraft flyable and operate them effectively, according to Syrian sources they were soon in action – and involved in several additional nocturnal reconnaissance sorties over Israel. (Photo by Erwin Hausmann)

instructional aid and spare parts for as long as the aircraft remained in service.[5]

In the meantime, ties to Prague were back to earlier levels, and in the course of several meetings through 1962 and early 1963, the two governments entered negotiations for the construction of an entirely new Air Force Academy at the Kweres AB. Sadly for them, the Czechoslovak economy proved unable to find suitable contractors, or make an attractive offer. Thus, the order eventually went to Poland, while a similar contract for construction of a brand-new T-4 Air Base – atop a disused airfield that served one of the pumping stations on the pipeline connecting Iraq with Tripoli in Lebanon – went to Bulgaria. Instead, the Czechoslovaks were contracted with partial reconstruction of Kweres, expansion of Dmeyr AB (and its protection from frequent floods by nearby streams), expansion of the military side of Nayrab International, and other, minor construction work.[6] Moreover, a group of 40 Syrian cadets was sent to Czechoslovakia for three years of pilot training, and a group of Czechoslovak advisors was dispatched to Syria to provide advanced flight training.

On the dissolution of the UAR, the SyAAF was left with a miscellany of 40 MiG-17Fs, some of which were originally delivered to Egypt, and wore four-digit serial numbers, and some of which were originally delivered to Syria, and wore two- or three-digit serial numbers. All wore black identification stripes around their rear fuselages and wingtips. While applying a third green star on the white field of the national markings, the Syrians retained whatever other insignia was already on their aircraft. These four MiG-17s were photographed during a parade in Damascus, in 1963 or 1964. (Nour Bardai Collection)

FATEFUL DMEYR AIR BASE

Another affair that remained constant during this period of Syrian history was the Israeli treatment of the country. With the Syrian armed forces in disarray, Israel felt free to not only finish the destruction of the construction site outside the village of Tewfik, but to entangle its north-eastern neighbour in a gradually intensifying campaign of provocations. The pattern of that operation was similar to the one from 1958-1960: the IDF would deploy armoured tractors to 'work the soil' inside the DMZ and regardless of whether the Syrians opened fire or not, the Israelis would fire at the nearest enemy positions. Whenever the Syrians dared fire back, the IDF would hit back with artillery and, later on, air power. This is what had happened during the first Israeli attack of this period, on 21 March 1962: when a Syrian outpost in the Nukhelia area opened fire at one of the tractors underway in the DMZ, it was raided by Israeli commandos, and when four SyAAF MiG-17s attempted to intervene, these were forced away by four SMB.2s, while still well inside Syrian airspace.[7]

While certainly satisfying for the Israelis, such provocations were 'no fun' for the Syrians, especially not for Qudsi's government that – despite its huge success on the economic level – found itself exposed to critique from almost every imaginable direction because of a failure to 'respond in kind'. Unsurprisingly, as soon as he was out of prison, Assad re-joined Jadid and Umran to – inspired by the apparent success of colleagues of the Ba'ath Party in Iraq – instigate a new coup d'état in Syria on 8 March 1963. While Jadid and Umran paralysed the command of the 70th Armoured Brigade in Damascus, Hafez al-Assad and his brother Rifaat secured the command of Dmeyr AB. The Ba'ath Party thus established itself in control over Syria. As a reward for his 'distinguished work and loyalty to the Party', Hafez was permitted to re-join the SyAAF with the rank of Captain and assigned the position of the Commander of the Fighter Force. Together with his brother, he then exploited the opportunity to establish the 'Defence Companies' – a de facto private militia – and to create a highly efficient intelligence network by appointing a number of confidants to sensitive positions in the military administration. This proved a timely decision, then on 18 July 1963, parts of the Army and Air Force launched a counter-coup, supported by four MiG-17 pilots from al-Mezzeh AB: while the leadership of the Ba'ath panicked, the intervention of Rifat's Defence Companies and paralysis of communication channels by Hafez's intelligence network managed to subdue the plot. Having proven his value to Jadid, less than a year later Hafez al-Assad was appointed the commander of the SyAAF.[8]

Meanwhile, the Israelis continued increasing their pressure. On 9 June 1963, the IDF opened fire at a small Syrian fishing village on the eastern side of Lake Tiberias and then subjected it to repeated air strikes. The reason was that the entire lake and exactly 30 feet (about 10 metres) of its eastern shoreline had been a part of the former British Mandate of Palestine, and thus came under Israeli control. Since 1949, the Syrians had thus required permission to fish in Lake Tiberias. Until 1963, nobody cared to ask nor to grant that permission, but now the IDF began shooting at the Syrian fishermen, insisting that Damascus must recognise Israel before any permission for fishery rights would be granted. Like all other Arab states of the time, Syria refused to recognise Israel because it occupied parts of Palestine not mandated to it by the UN Partition Plan of 1948 and expulsed the majority of the population.[9]

While UN observers managed to quieten the Syrians on that occasion, nothing similar was possible on 23 July 1963 when several Israeli fighter jets violated the airspace over the Golan Heights. This time, Damascus referred the situation all the way up to the Security Council of the UN: with there being no immediate reaction from any of the superpowers, the Israelis subsequently felt free to fly as deep inside Syrian airspace as they wanted. For example, late in the afternoon of 20 August 1963, four brand-new French-made Dassault Mirage IIICJ supersonic interceptors attacked four MiG-17Fs over the Golan Heights and claimed two as shot down. While the Syrians are not known to have suffered any such losses – and gunsights of Israeli Mirages at this point in time were anything but reliable or well-synchronised with their 30mm DEFA cannons – it became clear that the Mirages were vastly superior to whatever the SyAAF was operating at the time, and Damascus was in urgent need of new fighter jets.

A NEW DAWN IN BAGHDAD

If Nasser, Amer and their aides proved successful in turning the mass of the Syrian armed forces and political elite against themselves, Qasim did at least as well in turning almost everybody in Iraq against himself. Certainly enough, the country did not fall apart, but the local branch of the Ba'ath Party attempted to assassinate him in 1959; he distanced himself from well-organised Communists, de facto ruined much of the economy, and then also found himself at odds with the Kurdish leader Mustafa Barzanji. After bribing several Kurdish rivals of Barzanji, Qasim unleashed an offensive against insurgents that were hiding in the mountains of northern Iraq. Undertaken in the best traditions of what the Iraqis had learned from the British, the operation was intensively supported by the IrAF. Indeed, carefully guided by airborne forward air controllers (FAC) that flew US-supplied Cessna O-1 Brid Dogs, and entirely unopposed even by groundfire, Furies, Hunters, Il-28s, MiG-17s, Venoms and even a few Vampires then bombed more than 500 villages, driving the Kurds into Turkey and Iran. Qasim rushed to announce a victory in putting down the 'Kurdish uprising': in reality, it was only the beginning of serious troubles for him.

Although successful, this operation proved highly unpopular within the Iraqi armed forces, where dissent was already widespread due to economic difficulties. Awqati attempted to satisfy almost endless demands for new and better arms by placing an order for 16 MiG-21F-13 supersonic interceptors with Moscow: the first group of Iraqi pilots and ground crews thus travelled to the USSR for conversion courses in late 1961, and the new jets were all in Iraq by June 1962, where they entered service with the newly established No. 11 Squadron. However, once again, this was not really what the armed forces demanded: just like No. 9 Squadron

The Egyptian military photographer Osman Mahmoud arrived in Baghdad in time to take a series of photographs showing the destruction at Rashid AB caused by air strikes flown by al-Windawy and as-Saydoon on 14 February 1963. This one shows the wreckage of two MiG-19Ss. (Photo by Osman Mahmoud, Via Nour Bardai)

A holed and burned-out MiG-19S of No. 9 Squadron, IrAF, seen at Rashid AB after the coup of 14 February 1963. Despite such scenes, the top Egyptian military commanders showed remarkably little interest in solving the problem of exposure of their aircraft to similar attacks: this was to have fatal consequences four years later. (Photo by Osman Mahmoud, via Nour Bardai)

A Bofors 40mm anti-aircraft gun of the Iraqi Army – the same as used to shoot down the MiG-17F piloted by Fahad Abd el-Haley as-Saydon, the only loyal pilot to scramble during the coup of 8 February 1963. (Albert Grandolini Collection)

before, the new unit foremost served the purpose of improving protection of the government in Baghdad. Unsurprisingly, when Barzanji regrouped his forces and launched a new offensive and in March 1962, he nearly destroyed the 2nd Division of the Iraqi Army. Certainly enough, the IrAF once again reacted with a vicious air policing campaign. However, while demoralised and dispersed, the insurgents hit back by blowing up bridges and roads. Through late 1962 and early 1963, the war settled into a brutal routine: Barzanji's forces controlled much of the countryside by night, while by day the IrAF bombed whatever it could find. Life in nearly a third of Iraq north of Mosul and Kirkuk was thus completely paralysed. Before long, the Iraqi Ba'ath Party was on the move: establishing contact with officers serving with Nos. 6 and 7 Squadrons, home-based at Habbaniya AB, it plotted a coup emphasising the destruction of Nos. 5, 9 and 11 Squadrons at Rashid AB, all of which were 'staffed by Communists protecting Qasim'. At dawn on 8 February 1963, elements of the 20th Brigade led by Colonel Abd as-Salam Arif seized Baghdad radio station, while two Hunters from No. 6 Squadron – flown by Munthir al-Windawy and Mumtaz as-Saydon – repeatedly attacked Rashid AB, destroying between four and six MiG-19s on the ground. Minutes later, a follow-up strike by two MiG-17Fs from No. 7 Squadron knocked out three MiG-21F-13s. The only MiG-17F that managed to scramble in response, piloted by Fahad Abd el-Haley as-Saydon from No. 5 Squadron, was then shot down by a battery of 40mm Bofors anti-aircraft guns deployed to protect Rashid AB. With the 'flying praetorian guard' out of the way, the Ba'ath militia and the Air Force then pushed to surround and isolate the government: Qasim and his aides thus found themselves locked up in the building of the Ministry of Defence. Over the following hours, this was subjected to up to 60 air strikes by additional Hunters and MiGs. After a 10-hour battle, the badly shaken Qasim gave up: he was arrested together with Awqati, and both were executed on 14 February 1963.[10]

The government that subsequently took over in Baghdad was composed of Ba'athist civilians and pan-Arab military officers and presided over by Arif (who promptly advanced himself in rank to the first Field Marshal in Iraq's history). One of its first decisions was to terminate all the military cooperation with the USSR. Immediately afterwards, the Ba'ath militia launched a merciless persecution of the Leftists. Unsurprisingly, the IrAF was hit hard by purges, having multiple squadron commanders and other ranks at least distanced from the service, if not imprisoned, while others fled to Jordan and Syria. Indeed, because all of the Soviet advisors were expulsed from Iraq, the Air Force abandoned the project of introducing to service four battalions of Soviet-made S-75 (ASCC/NATO-codename 'SA-2 Guideline') surface-to-air missiles (SAMs) delivered in 1961. These were returned to Moscow in mid-1963, and the Iraqi personnel undergoing training in the USSR were recalled. The IrAF thus missed the opportunity to become the first air force in the Middle East to operate SAMs.[11]

Thus ended the first phase in which Egypt and Syria, followed by Iraq, began acquiring military aircraft of Soviet origin. It was a colourful and exciting period of rapid developments and growth, which can be summarised as one with far-reaching consequences – for the air forces involved, for many of their pilots, and for entire nations. However, even more massive adventures were waiting for both the aircraft in question and their pilots and ground crews: that story is to be told in Volume 2.

Table 6: Known Iraqi Acquisitions of Soviet-made Military Aircraft, 1958-1963[12]

Aircraft Type	Number
An-2	4
Il-28	20
Il-28BM	2
Il-28U	2
Mi-4	13
MiG-17F	11
MiG-17PF	17
MiG-19S	16
MiG-21F-13	16
Tu-16	10
Yak-18	10

BIBLIOGRAPHY

PRIMARY DOCUMENTS

Manuscripts and photographs from private documentation of the following veteran EAF, IrAF, QJJ, and SyAAF pilots were used:

- Mohammad Ahmad (IrAF)
- Fayez Baqir (IrAF)
- Rabee' Dulaymi (IrAF)
- Fikry el-Gindy (EAF/UARAF)
- Gabr Ali Gabr (EAF/UARAF; 'Notes')
- Magd el-Din Rifaat (EAF/UARAF; 'The Diary of a Mirage Pilot in the October War', 'Memoirs of Major-General Magd el-Din Rifaat)
- Jameel Salwan (IrAF)
- Mohammad Saaydon (IrAF; 'Pilot Memoir', privately published document from 2005)
- Mohammad Salman (IrAF)
- Haytham Khattab Omar (IrAF; 'Memoirs of the Commander of the Iraqi Air Force')

Air Ministry (UK), files as listed in endnotes
British Defence Attaché in Iraq (UK), files as listed in endnotes
CIA, *Iraq's Air Force: Improving Capabilities, Ineffective Strategy; An Intelligence Assessment*, October 1987, CIA/FOIA/ERR
DIA, *Electronic Warfare Forces Study – Iraq*, 9 August 1990, National Archives
Foreign Office (UK), files as listed in endnotes
Foreign Technologies Division (USAF), *Fishbed C/E Aerial Tactics* (tactical manual for MiG-21F-13, MiG-21PF, MiG-21FL and MiG-21PFM, obtained from Iraq in 1963 and translated to English by the Foreign Technologies Division USAF, in 1964)
Iraqi Air Force & Air Defence Command, *An Analytical Study on the Causes of Iraqi Aircraft Attrition during the Iran-Iraq War* (in Arabic), (self-published for internal use, May 1991; English transcription provided by Sadik)
Iraqi Air Force & Air Defence Command, *The Role of the Air Force and Air Defence in the Mother of All Battles: After Action Report* (in Arabic), (self-published for internal use, 5 November 1991, captured in 2003 and translated as '*A 1991 Dossier on the Role of the Iraqi Air Force in the Gulf War*', by the US Department of Defence-sponsored Conflict Records Research Center (CRRC Record Number SH-AADF-D-000-396) in the course of the 'Project Harmony')
Iraqi Air Force Martyrs Website, 1931-2003, *iraqiairforcememorial.com*
Iraqi Ministry of Foreign Affairs, *Iraqi Air Force Aircraft flown to Iran, in 1991* (in Arabic), (letter to the General Secretary of the UN, September 1991)
IRIAF, *204 KIA and 58 POW pilots of the Sacred Defence* (in Farsi), (self-published for internal use, listing 204 IRIAF pilots and crewmembers that were killed and 58 that were captured during the war with Iraq by their full rank and name, aircraft they flew, date of their death or captivity; date and place of issue unknown; copy provided by Farzin Nadimi)
Ministry of Defence (UK), files as listed in endnotes
Office of Naval Intelligence (USA), files as listed in endnotes

INTERVIEWS

As well as the officers mentioned in the acknowledgements, the following officers and other ranks of the SyAAF have been interviewed for this project over time, whose names we do not feel free to reveal for concerns related to their safety and the safety of their families:

'Adad' (retired Su-7/20/22-pilot, SyAAF), interview, May 1996
'Boudros' (retired MiG-21-pilot, SyAAF), interviews in March 2007, October 2008, November 2015, February 2016, April 2018
'Hussam' (retired MiG-21- and MiG-25-pilot, SyAAF), interview, October 2016, April 2018
'Duha' (retired SyAADF-officer), interviews, August 2004, March 2007, July 2013, November 2015, February 2016, April 2018
'E. R.' (veteran of the IDF), interviews, July 2012
'Hashim' (retired SyAADF-officer), interviews, March 2005, March 2006, July 2013, November 2016
'Ismael' (retired SyAAF MiG-21/Su-22/Su-24-pilot), interview, November 2015
'Jabbar' (retired officer of the Syrian Military intelligence), interviews, June 2003 and March 2006
'Nabil' (SA-6/11/17-operator, SyAADF), interviews, April 2001, March 2007, October 2007, July 2013, November 2016
'Mouhannad' (MiG-21- and Su-22-pilot, SyAAF), interview, August 2004, June 2011, July 2013, December 2016, April 2018
'Talal' (CO SAM-brigade, SyAAF/SyAADF), interviews, April 2001, August 2004, March 2005, March 2006, March 2007, October 2007, October 2008, December 2012, July 2013, November 2015, December 2016, April 2018

BIBLIOGRAPHY

Allawi, A. A., *Faisal I of Iraq* (Yale: Yale University Press, 2014)
Alnasrawi, A., *The Economy of Iraq: Oil, Wars, Destruction of Development and Prospects, 1950-2010* (Westport: Greenwood Press, 1994)
Ameri, H. A. 'Water War in the Middle East: A Looming Threat', *The Geographical Journal*, No. 168
Andersson, L., *Svenskarna I Jemen* (De Tog Flyget Till Medeltiden: Svenskarna i Jemen) (Uppsala: Z-förlaget, 2008)
Andrew, C. M., *France Overseas: The Great War and the Climax of French Imperial Expansion* (London: Thames and Hudson, 1981)
Asher, D., *The Egyptian Strategy for the Yom Kippur War: An Analysis* (McFarland & Co. Inc., 2009)
Awardi, A al-, *History of the Iraqi Armed Forces, Part 17: Establishment of the Iraqi Air Forces and its Development* (in Arabic, translation by Ali Tobchi) (Baghdad: Ministry of Defence, 1988)
Badri, H. el-, Magdoub, T. el-, Zohdy, M. D. el-Din, *The Ramadan War, 1973* (Dunn Loring: T. N. Dupuy Associates, 1978)
Bar-Noi, U., *Revelations From the Polish Archives*, Cold War International History Project at the Woodrow Wilson International Center for Scholars, e-Dossier No. 8 (based on 'On Soviet Policy Following the Israeli Aggression in the Middle East', by Comrade L. I. Brezhnev to the Plenum of the Central Committee of the CPSU, 20 June 1967, Archiwum Akt Nowych – a Polish document describing the speech given by Brezhnev to the Plenum of the Central Committee of the Communist Party of the Soviet Union on the actions undertaken by the Soviet leadership before and during the 1967 Arab-Israeli War).
Barr, J., *A Line in the Sand: Britain, France and the Struggle that shaped the Middle East* (London: Simon & Schuster UK Ltd., 2011)

Bull, General O., *War and Peace in the Middle East: The Experiences and Views of a UN Observer* (London: Leo Cooper Ltd, 1976)

Burns, W., *Economic Aid and American Policy towards Egypt* (New York: State University of New York Press, 1985)

Centre for Military Studies, *The History of the Syrian Army* (in Arabic), (Damascus: Centre for Military Studies, 2001-2002)

Cobban, H., *The Palestinian Liberation Organisation: People, Power, and Politics* (Cambridge: Cambridge University Press, 2008)

Cooley, J. K., 'The War over Water', *Foreign Policy* (Spring 1984), p.3-26

Cooper, T., *Syrian Conflagration: The Syrian Civil War 2011-2013* (Solihull: Helion & Co., 2015)

Cooper, T. & Salti, P., *Hawker Hunters at War: Iraq and Jordan, 1958-1967* (Solihull: Helion & Co., 2016)

Cooper, T., *Hot Skies over Yemen, Volume 1: Aerial Warfare over the Southern Arabian Peninsula, 1962-1994* (Solihull: Helion & Co., 2017)

Cooper, T., *MiG-23 Flogger in the Middle East: Mikoyan i Gurevich MiG-23 in Service in Algeria, Egypt, Iraq, Libya and Syria, 1973-2018* (Warwick: Helion & Co., 2018)

Cooper, T., *Moscow's Game of Poker: Russian Military Intervention in Syria, 2015-2017* (Warwick: Helion & Co., 2018)

Cooper, T. & Santana, S., *Lebanese Civil War, Volume 1: Palestinian Diaspora, Syrian and Israeli Interventions, 1970-1978* (Warwick: Helion & Co., 2019)

Cooper, T. & Sipos, M., *Iraqi Mirages: The Dassault Mirage Family in Service with the Iraqi Air Force, 1981-1988* (Warwick: Helion & Co., 2019)

Crosbie, S., *A Tactic Alliance: France and Israel from Suez to the Six Day War* (Princeton: Princeton University Press, 1974)

Cull, B., Nicolle, D., and Aloni, S., *Wings over Suez* (London: Grub Street, 1996)

Curtis, M., *Unpeople: Britain's Secret Human Right Abuses* (Vintage Books, 2004)

Dahl, B., *The Lebanese-Palestinian Conflict in 1973: The Social (De) Construction of Lebanese Sovereignty* (Thesis)(Oxford: University of Oxford, Faculty of Oriental Studies, 2006)

Dawisha, A. I., 'Syria in Lebanon: Assad's Vietnam?', *Foreign Policy No. 33* (Winter 1978-1979)

Denis, A., *Historique de l'escadron de chasse 1-3 Navarre: 1915-2000* (A Denis, 2001)

Dimitrijevic, B., *UNEF: The Yugoslav Contingent; Yugoslav Army Contingent in the Sinai Peninsula, 1956-1967* (Warwick: Helion & Co., 2019)

Dresch, P., *A History of Modern Yemen* (Cambridge: Cambridge University Press, 2010)

Dupuy, Col. T. N. & Blanchard, W., *The Almanac of World Military Power* (New York: T. N. Dupuy Associates, 1972)

Eden, A., *Full Circle: The Memoirs of Anthony Eden* (Boston: Houghton Mifflin, 1960)

Even, Y., 'Two Squadrons and their Pilots: The First Syrian Request for the Deployment of Soviet Military Forces on its Territory, 1956', Cold War International History Project, Working Paper No. 77, Woodrow Wilson International Center for Scholars, February 2016

Fawzy, Major-General M., *The Three-Years War* (in Arabic) (Beirut: Dar Mustakbal al-Arabi, 1998)

Ferris, J., Soviet Support for Egypt's Intervention in Yemen, 1962-1963, *Journal of Cold War Studies*, Volume 10, Number 4, Fall 2008, pp.5-36 (article)

Flintham, V., *Air Wars and Aircraft: A Detailed Record of Air Combat 1945 to the Present* (London: Arms and Armour Press, 1989)

Gamasy, FM M., A., G., el-, *The October War: Memoirs of Field Marshal el-Gamasy of Egypt* (Cairo: The American University of Cairo Press, 1993).

Green, S., *Taking Sides: America's Secret Relations with a Militant Israel, 1948/1967* (London: Faber and Faber Ltd, 1984)

Green, S., *Living by the Sword: America and Israel in the Middle East* (Brottleboro: Amana Books, 1988)

Green, W., and Fricker, J., *The Air Forces of the World* (London: MacDonald, 1958)

Gindy, F. el-, *Egyptian Eagles over the Golan* (in Arabic) (Cairo: al-Hay'ah al-Misriyah al-'Ammah lil-Kitab, 1992)

Gordon, S., *Thirty Hours in October* (in Hebrew) (Tel Aviv: Ma'ariv Book Guild, 2008)

Group of Authors, *The Role of the Iraqi Armed Forces in the October 1973 War* (in Arabic) (Beirut: Establishment for Arab Studies and Publication, 1974)

Gunston, B., *An Illustrated Guide to Modern Fighters and Attack Aircraft* (London: Salamander Books Ltd., 1980)

Gunston, B. & Spick, M., *Modern Air Combat: The Aircraft, Tactics and Weapons Employed in Aerial Warfare Today* (London: Salamander Books Ltd., 1983)

Gurov, I. M., *Ground-based Air Defence Systems at Wars in Vietnam and the Middle East* (published in Russian for internal use as a training tool) (Moscow: Ministry of Defence of the USSR, 1980)

Haloutz, D., *Straightforward* (in Hebrew) (Miskal: Yedioth Ahoronoth Books and Chemed Books, 2010)

Hammad, G., *Military Battles on the Egyptian Front 1973* (in Arabic) (Cairo: Dar al-Shuruq, 2002)

Heykal, H. M. *The 30 Years War* (in Arabic) (Cairo: el-Ahram Publishing & Translation Centre, 1990)

Hoyt, T. D., *Military Industry and Regional Defence Policy; India, Iraq and Israel* (Oxon: Routledge, 2007).

Hughes, M., 'The Banality and Brutality: British Armed Forces and the Repression of the Arab Revolt in Palestine, 1936-1939', *English Historical Review*, Vol. CXXIV, No. 507.

International Center for Transnational Justice, *Lebanon's Legacy of Political Violence: A Mapping of Serious Violations of International Human Rights and Humanitarian Law in Lebanon, 1975-2008*, September 2013.

Jawadi, Dr. M. al-, *In Between the Catastrophe: Memoirs of Egyptian Military Commanders from 1967 to 1972* (in Arabic) (Cairo: Dar al-Khiyal, 2001)

Khalidi, R., *The Hundred Years' War on Palestine: A History of Settler Colonial Conquest and Resistance, 1917-2017* (Profile Books, 2020)

Konzelmann, G., *Damaskus: Oase Zwischen Hass und Hoffnung* (Frankfurt/Main: Ullstein Buch, 1996)

Kotlobovskiy, A. B., *MiG-21 in Local Wars* (in Russian) (Kiev, ArchivPress, 1997)

Langeron, J.-M., *F-4 Phantom II Flight Model Identification & Compared Air Combat Performance Analysis: MiG-21 versus F-4 Phantom II* (Parts 1, 2 and 3). These privately published assessments are based on following documents:

- *Flight Manual F-4C,/D,E-1-S, T.O. 1F-4C-1*, 15th August 1973, Change 1
- *Performance Data Manual F-4C, D, E-1-1-S, T.O. 1F-4C-1-1*, 15 July 1969, Change 1
- *Flight Manual F-4E, T.O. 1F-4E*, 1 February 1979

- *NATOPS Flight Manual MiG-21M*
- *NATOPS Flight Manual MiG-21MF*
- *NATOPS Flight Manual F-4D blk37*
- *NATOPS Flight Manual F-4E blk31*
- *NATOPS Flight Manual F-4E blk41*
- *NATOPS Fight Manual F-4E blk46 To.556*
- *NATOPS Flight Manual F-4E blk50*
- Robert K Heffley & Wayne F Jewel, NASA CR-2144: *Aircraft Handling Qualities Data*, NASA, December 1972
- Joseph R Chambers & Ernie L Anglin, NASA TN D-5361: *Analysis of lateral-directional stability characteristics of a twin-jet airplane at high angles of attack*, NASA, August 1969.

Laron, G., 'Cutting the Gordian Knot: The Post-WWII Egyptian Quest for Arms and the 1955 Czechoslovak Arms Deal', Cold War International History Project, Working Paper No. 55, Woodrow Wilson International Center for Scholars, February 2007.

Marashi, I. al- & Salama, S., *Iraq's Armed Forces: An Analytical History* (Abingdon: Routledge Middle Eastern Military Studies, 2008)

Melman, Y., *The Master Terrorist: The True Story Behind Abu Nidal* (London: Sidgwick & Jackson, 1987)

Menahim, Maj Gen K. al-, *Egyptian Wars in Modern History* (in Arabic) (Beirut: Dar Mustakbal al-Arabi, 1990)

Meyer, K. E. & Brysac, S. B., *Kingmakers: The Invention of the Modern Middle East* (New York: W. W. Norton & Co. 2009)

Michel, M. L., *Clashes: Air Combat over North Vietnam, 1965-1972* (Annapolis: Naval Institute Press, 1997)

Moneim el-, A. A., *Wolf in the Sun's Disc* (in Arabic) (Cairo, self-published, 1988)

Morris, B., *Israel's Border Wars, 1949-1956: Arab Infiltration, Israeli Retaliation, and the Countdown to the Suez War* (Oxford: Oxford University Press)

Morris, B., *Righteous Victims: A History of the Zionist-Arab Conflict, 1881-1999* (London: Random House, 1999)

Morris, B., *The Birth of the Palestinian Refugee Problem, 1947-1949* (Cambridge: Cambridge Middle East Library, 2004)

Moukiiad, Major-General M. A., *My Life* (in Arabic) (Damascus: az-Zakhira, 2005)

Mustafa, General H., *The June War, 1967, Part II* (in Arabic) (Beirut: Establishment for War Studies, 1970)

Nicolle, Dr. D., Gabr, Air Vice-Marshal G. A., *Air Power and the Arab World, Volume 1, Military Flying Services in the Arab Countries, 1909-1955* (Warwick: Helion & Co., 2019)

Nicolle, Dr. D. & Gabr, Air Vice-Marshal G. A., *Air Power and the Arab World, Volume 2, Military Flying Services in the Arab Countries, 1909-1955* (Warwick: Helion & Co., 2020)

Nicolle, Dr. D. & Gabr, Air Vice-Marshal G. A., *Air Power and the Arab World, Volume 2, Military Flying Services in the Arab Countries, 1909-1955* (Warwick: Helion & Co., 2020)

O'Ballance, E., *The Kurdish Revolt 1961-1970* (London: Faber and Faber Ltd., 1973)

O'Ballance, E., *No Victor, No Vanquished: The Yom Kippur War* (Novato: Presidio Press, 1996)

Okasha, Maj Gen Mohammed Abdel Moneim Zaki, *Conflict in the Sky: the Egyptian-Israeli Wars, 1948-1967* (Cairo: Ministry of Defence, 1976)

Okasha, Maj Gen Mohammed Abdel Moneim Zaki, *Soldiers in the Sky* (in Arabic) (Cairo: Ministry of Defence, 1976)

Okasha, Maj Gen Mohammed Abdel Moneim Zaki, *Operation Spark: I Wave* (in Arabic) (Cairo: Ministry of Defence, 2011)

Okasha, Maj Gen Mohammed Abdel Moneim Zaki, *69 Squadron at War* (in Arabic) (Cairo: Ministry of Defence, 2012)

Ovendale, R., *The Origins of the Arab-Israeli Wars* (Harlow: Longman Group UK Ltd, 1984)

Pappe, I., *The Ethnic Cleansing of Palestine* (Oxford: Oneworld Publications, 2006)

Peled, Gen. B., *Days of Reckoning* (In Hebrew) (Moshav Ben-Shemen: Modan Publishing House, 2004)

Pivka, O. von, *Armies of the Middle East* (London: Book Club Associates, 1979)

Provence, M., *The Great Syrian Revolt and the Rise of Arab Nationalism* (Austin: University of Texas Press, 2005)

Provence, M., *The Last Ottoman Generation and the Making of the Modern Middle East* (Cambridge: Cambridge University Press, 2017)

Raspletin, Dr. A. A., 'History PVO' (website in Russian: *historykpvo.narod2.ru*), 2013.

Riad, M., *The Struggle for Peace in the Middle East* (Consett: Quartet Books, 1981)

Safran, N., *From War to War: the Arab-Israeli Confrontation*, 1948–1967 (New York, Pegasus Books, 1969)

Sampson, A., *Die Waffenhändler: Von Krupp bis Lockheed, Die Geschichte eines tödlichen Geschäfts* (Reinbek bei Hamburg: Rowohl Verlag GmbH, 1977)

Schmidt, D. A., *Yemen: The Unknown War* (London: The Bodley Head Ltd. 1968)

Segev, T., *1967: Israel, the War, and the Year that Transformed the Middle East* (Abacus, 2008)

Shalom, D., *Like a Bolt out of the Blue: 'Moked' Operation in the Six Day War, June 1967* (in Hebrew) (Rishon Le-Zion: BAVIR Aviation Publications, 2002)

Shazly, Sa'ad el-, *The Crossing of the Suez* (San Francisco: American Mideast Research, 2003)

Shevchuk, D., *The Soviet-Israeli War* (in Russian; unpublished manuscript)

Shlaim, A., *The Iron Wall: Israel and the Arab World* (W. W. Norton & Company, 2001)

Shlaim, A., *Lion of Jordan: The Life of King Hussein in War and Peace* (New York: Vintage Books, 2009)

Shukairy, A., *The Great Defeat; Major Defeat of Kings and Presidents* (Cairo: Arab Foundation for Publishing and Distribution, 2005)

Sipos, M. & Cooper, T., *Wings of Iraq, Volume 1: The Iraqi Air Force, 1931-1970* (Warwick: Helion & Co., 2020)

Sokolov, A, 'PVO in Local Wars and Armed Conflicts: The Arab-Israeli Wars', *VKO*, No. 2 (2), 2001

Sokolov, A., 'The Arab-Israeli Wars', *VKO*, No. 2 (5), 2002

Sorby, Jr., K., '1941 British-Iraqi War', *Military History* (in Slovakian), 3 December 2008

Sorby, Jr., K., 'Approach of the Ba'ath Party to the Kurdish question in Iraq After 1968', *Asian and African Studies*, Volume 27, Number 2, 2018

Sorby, Jr., K., 'Iraq 1963: The Short Rule of the Ba'ath', *Asian and African Studies*, Vol. 18, 2009

Sorby, Jr., K., 'Iraq Under the Reign of Faysal II (1953-1958)', *Asian and African Studies*, Vol. 13, 2004

Sorby, Jr., K., 'The 1952 Uprising in Iraq and Regent's Role in its Crushing: Iraq from al-Watba to al-Intifada, 1949-1952', *Asian and African Studies*, Vol. 12, 2003

Sorby, Jr., K., 'The Army's Growing Influence in Iraqi Politics (1937-1939)', *Yearbook of Oriental Studies* (in Slovak), T. LXVIII, Z. i. 2015, 20-36, (Bratislava, 2015)

Sorby, Jr., K., *Iraq Under the Rule of Abdassalam Arif (1963-1966)* (Bratislava: Institute of Oriental Studies, Slovak Academy of Sciences, 2010)

Sorby, Jr., K., *Iraq: The Rule of Abdarrahman Arif and Its End (1966-1968)* (Bratislava: Institute of Oriental Studies, Slovak Academy of Sciences, 2011)

Sorby, Jr., K., *The Two July 1968 Coups in Iraq: the Ba'ath is returning to Power* (Bratislava: Institute of Oriental Studies, Slovak Academy of Sciences, 2012)

Stafrace, C., *Arab Air Forces* (Carrolton: Squadron/Signal Publications Inc., 1994)

Tessler, M. A., *A History of the Israeli-Palestinian Conflict* (Bloomington and Indianapolis: Indiana University Press, 1994)

Thompson, Sir R. E., (editor), *War in Peace: An Analysis of Warfare since 1945* (London: Orbis Publishing, 1981)

Thornhill, 'Britain, the United States and the Rise of an Egyptian Leader', *English Historical Review*, Volume CXIV, Issue 483, September 2004, pp.893-894

Transue, J. R. (project leader), *WSEG Report 249: Assessment of the Weapons and Tactics used in the October 1973 Middle East War*, (Arlington: Institute for Defense Analyses/Center for Naval Analyses, Weapons Systems Evaluation Group, October 1974, CIA/FOIA/ERR)

Tripp, C., *A History of Iraq* (Cambridge: Cambridge University Press, 2000)

Van Creveld, M., *The Sword and the Olive* (New York: BBS Public Affairs, 1998)

Weiss, R. & Aloni, S., *Dassault Mystère IV* (Kefar-Tavor: IsraDecal Publications, 2010)

Williams, A. G. & Gustin, Dr. E., *Flying Guns: The Modern Era* (Ramsbury: The Crowood Press Ltd., 2004)

Zidek, P., Sieber, K., *Czechoslovakia in Middle East of the Years 1948-1989* (in Czech) (Prague: Ustav mezinarodnich vztahu, 2009)

Zidek, P., 'Vyvoz zbrani z Ceskoslovenska do zemi tretiho sveta, 1948-1962', *Historie a vojenstvi*, 3/2002

Zaloga, S., *Scud Ballistic Missile and Launch Systems, 1955-2005* (Oxford: Osprey Publishing Ltd, 2006)

Zaloga, S., *Red SAM: The SA-2 Guideline Anti-Aircraft Missile* (Oxford: Osprey Publishing Ltd, 2007)

Zolotaryov, Maj Gen V. A., *Russia in Local Wars and Military Conflicts in the Second Half of the 20th Century* (in Russian) (Moscow: Institute of Military History, Ministry of Defence of the Russian Federation, 2000)

Various volumes of *Armed Forces Magazine*, published by the Egyptian Ministry of Defence, 1950s and 1960s; *El-Djeich* (official publication of the Algerian Ministry of Defence), various volumes from 2007 to 2009; various magazines and journals published by the Iraqi Air Force and the Iraqi Ministry of Defence, 1970s, 1980s and 1990s; *Kanatlar* magazine (Turkey), June 2003.

NOTES

CHAPTER 1

1 The use of the expression 'colonisation of Palestine by Jewish immigrants from Europe' might appear aggressive, openly taking sides, or even anti-Semitic to many readers. For the reasons why this term was used, including very detailed descriptions of the affairs in question, see works such as those by Khalidi, Morris, and Pappe (details in bibliography).

2 For a detailed reconstruction and sources on all the above-listed developments, see Nicolle et al, *Wings over Sinai*.

3 Wagdi Hafez, interview, 11/2015.

4 CIA, 'Syria's Elite Military Units: Key to Stability and Succession: An Intelligence Assessment', NESA 87-10012, February 1987, CIA/FOIA/ERR.

5 CIA, 'Background Information on German Military Experts in Syria', 22 January 1954, CIA/FOIA/ERR. Notably, at the high point of the Axis advance into Egypt, in the summer of 1942, Rauff was deployed to North Africa with the task of organising the persecution of Jews in Egypt and Palestine once these had been captured. When the Axis advance was stopped at El Alamein, and then forced into a retreat, Rauff's *Einsatzkommando* was deployed in Tunisia instead.

6 Dukovac and his colleagues had flown for the quisling Croatian Air Legion (Kroatische Luftwaffen Legion) – a unit equipped with German aircraft and deployed on the Eastern Front against the USSR, during the Second World War. According to Jelavic (*352 [Y] RAF Squadron*), Dukovac and his colleagues defected to the Soviet side in 1944 and subsequently joined the emerging Yugoslav Air Force. Exposed to frequent abuse by Serbian officers, they then defected to Italy and were interned. Immediately following the armistice between Israel and Syria, the IDF contracted the same Croatian team – minus Dukovac: he had emigrated to Canada. However, his colleagues served as instructors for Israeli pilots before returning to Italy in late 1950.

CHAPTER 2

1 Unless stated otherwise, the following chapter is based on Nicolle et al, *Wings over Sinai*, and Smisek, *Czechoslovak Arms Exports in the Middle East, Vol.1*, Chapter 3 & *Vol.2*, Chapter 1.

2 Gabr Ali Gabr, *Notes* & Laron, pp.27-28; Gabr's notes contain an observation about reports from the British intelligence related to Mystère IVAs Israel ordered during the summer of 1955. This not only led to an assessment that such aircraft would provide Israel with air superiority throughout the Middle East, but also caused the British to worry about Israel planning an invasion of Jordan. Accordingly, this would have further strengthened Nasser's determination to purchase Soviet arms.

3 Smisek, *Czechoslovak Arms Exports, Vol.2*, Chapter 1.

4 Cull et al, p.86. According to Israeli sources, one of the MiG-15s involved in the first public appearance of the type in Egypt crashed while returning to base, see Even, p.6.

5 *Memorandum of Conversation at the White House*, from John Foster Dulles, 11 January 1956, Papers of John Foster Dulles, Box 10, Dwight D Eisenhower Library.

6 Morris, p.291.

7 Smisek, *Czechoslovak Arms Exports, Vol.1*, Chapter 3 & *Vol.2*, Chapter 1.

8 Fullick & Powell, p.11.

9 Green, p.123.
10 Even, p.9.
11 Cull et al, p.68.

CHAPTER 3

1 Gabr Ali Gabr, *Notes.*
2 Smisek, *Czechoslovak Arms Exports, Vol.2*, Chapter 3.
3 Paul Gaujac, *Suez 1956* (Paris: Facon, 1986); Smisek, *Czechoslovak Arms Exports, Vol.2*, Chapter 3 & information from the EAF Historical Department provided by Sherif Sharmi.
4 In his Notes, Gabr Ali Gabr stressed that the EAF was fully aware of overflights by British reconnaissance aircraft in September and early October 1956 but had clear instructions from Nasser not to interfere with them in order not to provoke a war, but also to avoid revealing its improved capabilities.
5 Smisek, *Czechoslovak Arms Exports, Vol.2*, Chapter 1.
6 Gabr Ali Gabr, *Notes*; Alaa Barakat, interview, 03/2003; 'Air Superiority and Airfield Attack: Lessons from History', BDM Corporation Report, prepared for the Defense Nuclear Agency, 1982; 'Operation Kadesh, IDF/AF 1950-1956, Buildup and Operations', IDF/AF History Branch, 1986. Such high numbers of available Egyptian pilots might appear surprising to many, especially considering the never-ending flow of reports in the Western media about the EAF lacking qualified fliers and technical personnel and reporting supposedly based on 'infallible Israeli intelligence'. However, they were confirmed by multiple additional sources, including statements by Nasser citing the deployment of 250-260 pilots for conversion purposes to Czechoslovakia and the Soviet Union in 1956, and documentation left behind by Abd el-Moneim el-Shennawy, who specified that after the Suez War he was sent to Czechoslovakia for a conversion course on MiG-15s together with 55 other pilots. The EAF could not have afforded to send so many pilots abroad for periods of between three and 12 months if it was as short on pilots as usually reported by Israel and the West.
7 Based on cross-examination of information from Gabr Ali Gabr, *Notes*; Barakat, interview, 03/2003; Labib, *The Third Arm* and information from the EAF Historical Department provided via Sherif Sharmi.
8 Even, p.9.
9 Even, p.7, citing V. A. Zolotarev, *Russia in Local Wars and Military Conflicts in the Second Half of the 20th Century* (in Russian), (Moscow, Kuchkovo Pole, 2000), p.174. Reports about deployment of pilots from the 119th Fighter Aviation Regiment (119. IAP) – including Anatolievich Sintsov, a veteran of the Second World War and Korean War – to Egypt, and them flying combat sorties, even shooting down the Westland Wyvern fighter-bomber flown by Lieutenant Dennis McCarthy of 802 Naval Air Squadron of the Royal Navy (McCarthy ejected safely), published in the Russian and Argentine press in the 1990s, all proved unsubstantiated. Certainly enough, Sintsov passed away in 1990, shortly after barely mentioning his three claims from the Korean War. Neither he nor any other Soviet pilot is known to have ever claimed any kind of combat operations in Egypt. Nevertheless, articles with such content were provided to Even by researchers, 'affiliated with the Institute for Military History of the Russian Federation's Ministry of Defence'. Correspondingly, these claimed that 'Soviet flight instructors fought wing-by-wing with Egyptian pilots'. One of them reported that 'Already at dawn of 30 October, they managed (with MiG-15s) to intercept four English Canberra espionage planes and shoot down one. On the next day, 31 October, Soviet pilots took part in attacking the outposts of (IDF) Brigade 202. On 1 November, a group of MiG-17 interceptor planes from the USSR joined, especially for the battle, and on 2 and 3 November managed to shoot down several British fighter planes'. Other similar reports from Russian sources cited Soviet pilots flying Egyptian Il-28s and participating in 'uncharacteristic' combat operations. In one case (at an undetermined date), three Soviet-flown Il-28s encountered '10 British aircraft over the suburbs of Cairo' and, supposedly, 'two Hunters were shot down' by the Ilyushins. Suffice to say that there is no trace of substance for any of these claims, and that the available Egyptian documentation stresses no Soviet involvement. Moreover, legends of such kind are not reserved for Russian sources alone. Doug Gordon wrote the following in the article 'USAFE's 38th TRS, 1952 to 1991', published in *AirEnthusiast* magazine, Volume July/August 2007:

> On the fifth day of our rest (after the gruelling 40-minute flight from Brindisi to Larissa), we were taken to the Athens Int'l Airport, to be picked up by Maj Rufus Barnes (10th Tactical Recon Wing Ops) in [a Douglas] RB-26. While we were waiting for be picked up, standing in the shade of the US Naval Attache's [Grumman] SA-16, 21 MiG-15s landed and were refuelled by Royal Dutch Petroleum's fuel trucks (with white sidewall tires, no doubt bought with US dollars…). The Russian pilots were wearing flight suits without insignia or other markings. After refuelling, they took off, bound for Cairo. Over the Med, British and French pilots, flying NATO fighters based on Cyprus, intercepted the Russians. Nineteen MiGs were shot down en route, one was shot down on final approach to Cairo Airport, and the other MiG landed and was strafed on the taxiway. While that was happening, we were making the turn over Marseille, heading home, when we got a radio call to proceed directly to Spangdahlem [NATO air base in then West Germany], and go directly to 38th TRS operations. (At that time, we did not know about the shoot-downs: we found that out after landing at Spang). For three days, 'the buzz' was that World War III was about to start. The Israeli Army had crossed the Suez, and was advancing towards Cairo. The Russian MiGs were supposed to assist the Egyptians, and counter the Israeli advance.

There remains neither evidence, nor a trace of reason for sighting '21 MiGs' at Athens International as of that time: Egyptian and Syrian jets were evacuated from Egypt during the Suez War, but to Saudi Arabia, not to Greece. Indeed, MiG-15s would have lacked the range to reach Athens, and one must wonder how the Egyptians would have managed to fly so many aircraft past six British and French aircraft carriers, and another of the US Navy, entirely unobserved. Indeed, the idea of the Egyptians – or Soviets – flying their aircraft over the Mediterranean Sea, where the sky was full of enemy aircraft, remains beyond the comprehension of the authors.
10 Interview and correspondence with Air Marshal Sa'ad ad-Din Sharif, acting ADC to President Sadat, 14 December 1973, Cairo (ad-Din Sherif had served in the RAF's No. 216 Transport and Ferry Group during the Second World War) & Ehud Younay, *No Margin for Error: The Making of the Israeli Air Force* (New York, 1993), pp.160-62.
11 According to Gabr Ali Gabr, *Notes*, while considering the French-Israeli collaboration as possible – especially in the light of exposed position of the Israeli paras the Heitan Defile – Nasser could not

bring himself to believe that the British would get involved on the side of the Israelis.

12 The aircraft in question might have been an Avro Lincoln of the secretive No. 199 (Electronic Countermeasures) Squadron, RAF.

13 Abd el-Moneim el-Tawil, interview, 02/1999; Sidqi, interview, 02/1999 & Smisek, *Czechoslovak Arms Exports, Vol.2*, Chapter 1.

14 Nour Bardai, interview, 10/2019.

15 Gabr Ali Gabr, *Notes*.

16 After spending several months as a prisoner of war in Israel, Ahmad Faraghal returned to Egypt and continued his career with the EAF as a flight instructor. That was the rule for all the ex-POWs serving with EAF, and it was changed only after the June 1967 War, when there was a great need for fighter pilots.

17 Authors' notes, based on an article by A. B. Kotlobovkiy, about Col. A. Bozhenko's tour of duty in Egypt, in 1956-1957, published in Ukrainian magazine Aviyatsya & Vremya, in the 1990s.

18 'The Wild Horses, the Mustang Squadrons', *Born in Battle* No. 44, 1987. Accordingly, a Mustang pilot '…K, forced down twice, by AA fire and an Egyptian Vampire, and thought dead, managed to hitch a ride back to Ekron and surprise his squadron commanders'. Cohen (pp.125-128) cites a loss under similar circumstances, on the same date and around the same time and place but caused by ground fire.

19 Farouq el-Ghazzawi, interview, 1997.

20 Fouad Kamal, interview, 02/1999.

21 Barakat, interview, 03/2003. Barakat's recollection about a highly effective air strike on the 202nd Brigade was confirmed by several other sources, but most of these placed it at the Mitla Defile, although the Israelis never came as far, and the column in question was attacked nearly 50 kilometres north-east of the strategic pass. For example, see Nordeen, p.43.

22 The French were not involved, as they did not have long-range bombers in service. Instead, their transport aircraft ferried supplies to Israeli units in Sinai.

23 Wagdi Hafez, interview, 11/2015.

24 Even, p.5, based on Air Intelligence Report No. 28, 11 November 1956, Air Force HQ, Air Branch 4, MD/6/2143, IDFA, file 675-535/2004. According to Even, '23 Syrian planes were damaged' during the Suez War. This number might include three out of four MiG-15UTIs that were eventually evacuated to Syria via Saudi Arabia and Jordan, and 20 MiG-15bis from the second batch ordered by Syria, that were not fully assembled and stored at Abu Suweir.

25 Denis Albin, *Historique de l'Escadron de Chasse 1/3 Navarre de 1915 à 2000* (relevant excerpt forwarded by a French reader of the *Arab MiGs* book-series).

26 This aircraft was donated to the SyAAF after the war.

27 Cull et al, pp.213 & 231.

28 Obviously not knowing about Nasser's order for the EAF to stop fighting, Air Marshal Denis Barnett, Air Task Force Commander of the RAF, stated that, '…if we had been up against an enemy with even a modicum of fighting qualities with the modern aircraft and equipment the Egyptians had, the situation would have been different…' (see Cull et al, p.198).

29 Cull et al, pp.250 & 260.

30 Cull, p.237.

31 Cull, pp.271 & 277.

32 Morris, *Righteous Victims*, pp.419-425.

33 Just the final British tally for EAF aircraft destroyed included a total of 91 MiGs, 11 Meteors, 30 Vampires, and 26 Il-28s. Furthermore, the Israelis claimed to have shot down four MiGs and four Vampires.

34 Even, p.5, based on Air Intelligence Report No. 28, 11 November 1956, Air Force HQ, Air Branch 4, MD/6/2143, IDFA, file 675-535/2004. In comparison, Czechoslovaks (Smisek, *Czechoslovak Arms Exports, Vol.1*, Chapter 3), reported availability of 14 airworthy MiGs (including eight repaired from combat damage), 4 undergoing repairs, and five waiting for spares as of January 1957 (for a total of 23).

35 Smisek, *Czechoslovak Arms Exports, Vol.2*, Chapter 1

36 Williams et al, p.79.

CHAPTER 4

1 Smisek, *Czechoslovak Arms Exports,* Vol.2.

2 Unless stated otherwise, all data related to el-Shennawy's career is based on interviews with Captain Tarek el-Shennawy, Abdel Moneim el-Shennawy's son, 09/2010. Further details from Fikry el-Gahramy and Abdel Moneim el-Tawil, who underwent the same MiG-15 conversion course in Czechoslovakia, in 1957-1958, interviews, 02/1999.

3 Ghazzawy, interview, 02/1999. One of the Syrian pilots sent on a conversion course to MiG-17s in the USSR was Hafez al-Assad (Konzelmann, *Damaskus*, p.259-270). Most of sources describe Assad as being trained on MiG-19s, although no aircraft of this type were ever ordered by Syria.

4 Sidqi, interview, 02/1999; Abd el-Moneim el-Tawil, interview, 02/1999; & Konzelmann, *Damaskus*, p.259-270. Notably, according to Konzelmann, Assad was assigned the command of a MiG-17 unit upon his return from the conversion course in the USSR and until the creation of the UARAF. Furthermore, although the UARAF introduced Arabic designations for military ranks on its establishment, the mass of contemporary Egyptian, Iraqi, and Syrian pilots preferred to use their 'old', RAF-style ranks.

5 Unless stated otherwise, this sub-chapter is based on Badr Domair, interivew, 07/2010 & Ahmed el-Keraidy, interview, 08/2010. Notably, Domair – who was personally involved in some of the operations in question – stressed that while several Syrian crews were converted to Il-28s by 1960, none was ever involved in overflights of Israel.

6 Unless stated otherwise, this and the following sub-chapter are based on Cooper et al, *Lebanese Civil War* & Sipos et al, *Wings of Iraq*.

7 For a detailed history of the Iraqi armed forces and the Iraqi air force up to 1970, see Sipos et al, *Wings of Iraq, Vol.1* – upon which, unless stated otherwise, this and the following sub-chapters are based.

8 'The King Chasers', *Time*, 24 November 1958. Eventually, UAR officials provided clear evidence that the flight crossed the border without proper clearance and had been intercepted by two MiGs in a routine and perfectly legal manner. Certainly enough, the Jordanian authorities had relayed the King's flight plan via Beirut to Damascus, as required by international aviation regulations, but nobody bothered to obtain the overflight clearance through diplomatic channels: instead, the Jordanians assumed that the flight had been 'cleared' through UN representative in Amman. Ultimately, the Jordanians realised that this was the responsibility of the commander of the RJAF, Lieutenant-Colonel Ibrahim Othman – who should have ordered 'someone' to obtain the clearance, but, still suffering from occasional blackouts from head injuries sustained when caught by an angry mob in Baghdad on 14 July 1958, failed to follow up – and thus fell silent in embarrassment. According to Patricia Salti, widow of a RJAF Hunter pilot and leading historian of that air force, both of the Syrian MiG-17 pilots defected to Jordan in May 1959 and requested political asylum.

9 For details on Israeli summary executions and atrocities against captured members of the Palestine Border Guard, the 8th (Palestinian) Division of the Egyptian Army, and the local civilians, during the occupation of November 1956 – March 1957, see Cobban, *The Palestinian Liberation Organisation*; Morris, *Israel's Border Wars*; Segev, *1967*; and Nicolle et al, *Wings over Sinai*. For a detailed coverage on one of the UNEF elements deployed on Sinai, see Dimitrijevic, *UNEF: The Yugoslav Contingent.*
10 Tawil, interview, 02/1999.
11 Badr Domair, interivew, 07/2010 & Ahmed el-Keraidy, interview, 08/2010.
12 "Chief of Staff from Major General A. R. Bolling", dated April 4, 1951, in Army Chief of Staff Decimal File, 1951-1953, File 092 Israel, Record Group 319, Records of the Army Staff, National Archives (USA).
13 Greene, *Taking Sides*, p.76-83.
14 Khalil, *The Fall of the Golan*. Al-Ja'una was an Arab village about 5km from Safad and near the Jewish town of Rosh Pinna, ethnically cleansed in mid-May 1948. Al-Khalisa was an Arab village about 28km north of Safad, ethnically cleansed after the latter town fell to the Haganah, on 11 May 1948: it was subsequently razed to the ground by the Israelis, who then constructed Kiryat Shemona partially atop its ruins. Kfar Shamir is the Syrian name for Kibbutz Shamir, established in north-eastern Upper Galilee by Jewish immigrants from Romania in 1944.
15 Gindy, interview, 02/1999 & Hafez, interview, 02/1999.
16 Based on documentation and interview provided by Patricia Salti, 11/2011.
17 Glassman, *Arms for the Arabs*, p.23.
18 Glassman, p.24; Samir Aziz Mikhail, interview, 06/2014; Barakat, interview, 03/2003.
19 Gamal Khalifa, interview, 03/2021
20 Unless stated otherwise, this sub-chapter is based on research for Smisek et al, *Wings of Iraq,* Vol.1.

CHAPTER 5

1 Keraidy, interview, 08/2010 & Boudros, interviews, 03/2007 & 10/2008.
2 Konzelmann, Damaskus, pp.259-270 & Sidqi, interview, 02/1999.
3 Sidqi, interview, 02/1999. His head injury ended Sidqi's highly successful flying career. Indeed, following his initial recovery, he was assigned as the Air Force Attaché to Switzerland, to be close to a suitable hospital.
4 Ajami, interview, 11/2006.
5 Moukiiad, *Autobiography*, 'Syria is Seeking Red Block's Help', *The New York Times*, 13 February 1962, and *The History of the Syrian Army*, Chapter 7.
6 Smisek, *Czechoslovak Arms Exports, Vol.1*, Chapter 3.
7 Moshe Dayan, interview with Rami Tal, 1976; this interview was published by *Jewish Chronicle* and *Yediot Achronot* only 20 years later, in May 1997, and quoted in the *New York Times* and *Associated Press* reports on 11 May 1997. Dayan clearly stated that Israel deliberately provoked attacks on Israeli farmers working below the Golan Heights as a pretext for the June 1967 War and the Israeli occupation of the Golan. "I know how at least 80% of the incidents began there. In my opinion, more than 80%, but let's talk about 80%. It would happen like this: We would send a tractor to plough some area of no value, inside the DMZ, and knew in advance that the Syrians would start to shoot. If they did not start shooting, we would tell the tractor to keep going forward, until the Syrians in the end would get nervous and start shooting. And then we would start firing artillery, and later also the air force, and this was the way it was. I did this, and Laskov and Tzur did it. Yitzhak Rabin did it when he was there [as commander of the Northern District IDF, in the early 1960s; authors' note], but it seems to me that it was Dado, more than anyone else, who enjoyed these games." Tal expressed his astonishment at these words, and asked what it was all for, to which Dayan explained: "What do I want to say by this? We thought then, and this continued for quite a long time, that we could change the lines of the armistice agreements by military actions that were less than war. That is, to grab some territory and to hang on to it until the enemy despairs and gives it to us." Dayan's daughter, Yael Dayen, later a Labour Member of the Knesset, Israel's Parliament, verified her father's remarks to Mr. Tal, who concluded, "We are not always the victims. There is a myth about peaceful farmers being shot at by Syrians – the truth is more ambiguous."
8 Moukiiad, *Autobiography* & Konzelmann, *Damaskus.*
9 Odd Bull, *War and Peace in the Middle East.*
10 Razzaq, interview, 04/2002; Sadik, interview, 03/2005; O'Ballance, pp.96-97 & Sorby (2009), pp.16-18.
11 Razzaq, interview, 04/2002; Sadik, interview, 03/2005; Awardi, p.77-80 & Sorby (2009), p.20.
12 Based on research for Sipos et al, *Wings of Iraq*; total count primarily based on cross-examination of available photographs.

ABOUT THE AUTHORS

Dr. David C Nicolle is a British historian specialising in the military history of the Middle Ages, with special interest in the Middle East and Arab countries. After working for BBC Arabic Service, he obtained his MA at SOAS, University of London, followed by a PhD at the University of Edinburgh. He then lectured in art history at Yarmouk University in Irbid, Jordan. Dr. Nicolle has published over 100 books about warfare ranging from Roman times to the 20th century, mostly as sole author. He also co-authored the 'Arab MiGs' series of books which covered the history of the Arab air forces at war with Israel from 1955 to 1973. Furthermore, he has appeared in several TV-documentaries, and has published numerous articles in specialised press. This is his fifth instalment for Helion's @ War series.

Tom Cooper is an Austrian aerial warfare analyst and historian. Following a career in the worldwide transportation business – during which he established a network of contacts in the Middle East and Africa – he moved into narrow-focus analysis and writing on small, little-known air forces and conflicts, about which he has collected extensive archives. This has resulted in specialisation in such Middle Eastern air forces as of those of Egypt, Iran, Iraq and Syria, plus various African and Asian air forces. In addition to authoring and co-authoring about 50 books – including about three dozen titles for Helion's @War series – and well over 1,000 articles. Cooper has been the editor of the five @ War series since 2017, and this is his 37th book for Helion.